HIVE MANAGEMENT

HIVE MANAGEMENT
A Seasonal Guide for Beekeepers

Richard E. Bonney

Storey Publishing

The mission of Storey Publishing is to serve our customers by publishing practical information that encourages personal independence in harmony with the environment.

Cover design by Cindy McFarland
Text design by Judy Eliason
Edited by Deborah Burns
Front cover photgraph by Stephan O. Muskie
Interior photographs by Richard Bonney unless otherwise noted
Illustrations by Alison Kolesar

The name Garden Way Publishing has been licensed to Storey Publishing, LLC, by Garden Way, Inc.

Printed in the United States by Versa Press
30 29 28 27 26 25 24 23

Library of Congress Cataloguing-in-Publication Data

Bonney, Richard E.
    Hive management: a seasonal guide for beekeepers / Richard E.
    Bonney
        p.   cm. — (Storey's down-to-earth guides)
    Includes bibliographica references (p.) and index.
    ISBN 978-0-88266-637-2
    1. Bee Culture. I. Title
SF523.B67   1991
638'.14 — dc20   90-55606
                                                              CIP

Contents

PREFACE

BEEKEEPERS TEND TO THINK OF BEEKEEPING and hive management only in terms of the current season — spring, summer, fall — whatever it might be. On the one hand, this is perfectly normal and natural; on the other hand, beekeeping over the year is an integrated activity, and actions taken in any given season should relate to both the past and the future. It can be difficult to correct a problem if too much time has passed since its inception, but with a little more attention to the colony during the previous season it might have been easy to anticipate the problem.

This book is intended to discuss hive management as an integrated activity, showing wherever possible how activities relate from season to season. It also is intended to explain the reasons for many of the practices that we follow without understanding why we do them. Because we don't know why, we often fail to take action at the proper time, or we take an action at an inappropriate time — reversing hive bodies too early in the season, for instance.

We should never forget that honey bees are wild creatures. They have never been domesticated. They have been kept, studied, researched, and bred for many years, and in a sense the species has been improved. But they have not been tamed. Left to their own devices they live exactly as they have lived for thousands of years. Our true, long-term success as beekeepers comes only after we have come to understand the intimate lives, behavior, and motivations of the bees. Otherwise, we are attempting to force a square peg into a round hole. Depending upon the size of the peg and of the hole we will sometimes have success, but seldom long-term success. To be successful it is imperative that the aspiring beekeeper have a good grasp of the behavior and motivations of the bee. Ultimately, these govern the life of the colony. To the extent possible this book will relate the motivations and behavior of the bee to management practices.

The beekeeping year is generally considered to begin in the spring. This is not unreasonable. It is the time that the bees become visibly active after the long winter. However, the normal colony is active during a good part of the winter, moving about within the hive, eating, raising brood, and even emerging on warmish days to fly. Because the bees are somewhat active during the winter, and because this activity, or lack of it, can have an effect on the coming season, we will consider the beekeeping season to begin in mid to late winter, in February or March. This is when the beekeeper should begin to think of the bees again and perhaps take some appropriate action.

This book then will loosely follow a beekeeping year that begins in February and continues until the end of the season when the bees have been put to bed for the winter.

CHAPTER 1

INITIAL INSPECTION

WE HAVE ALL EXPERIENCED one of those unseasonably warm days in February, just at that time when winter is weighing so heavily on us. Though snow may still be on the ground, the sun is shining, the temperature is in the sixties, and the bees are flying, a foretaste of spring to come. At least one of these days seems to occur every year. This is an excellent time to check on the bees, ensuring that all is well in anticipation of the hard days yet to come.

And the hard days will come. Just as we can expect one nice day in February, we can expect many harsh and blustery days in March and sometimes beyond. While browsing through an almanac recently I came across some old farmer's wisdom that seems to apply here. On the first of February, the almanac said, a farmer should still have on hand half of his winter's firewood and half of his winter's hay. This can very reasonably be extended to beekeepers and bees. On the first of February a colony should have on hand one half its winter's stores. For the bees as well as for the farmer it is a long time from the first of February until the start of the new growing season.

In February we begin to anticipate spring, and spring means bees flying regularly, foraging, expanding their population, and generally living a very active life. What is too often ignored is that this active life starts before spring. To most of us, spring really begins when the grass greens up and when flowers and leaves begin to appear, usually well into April. Even though they may not leave the hive, the bees anticipate spring well before this. During several weeks in the late fall and early winter the queen lays no eggs, but by February she is at it again, laying perhaps several hundred eggs per day. Consumption of honey and pollen increases rapidly as brood rearing picks up. If stores are low, of course, brood rearing is correspondingly low. An increasing rate of brood rearing as late winter and spring move along is critical for the later success of the colony. The population must be well above the winter low and rising when the serious nectar flows get underway in the late spring.

What is the beekeeper's role at this time? It is, as always, to ensure the well-being of the colony. First, recognize that every winter is different. Warmer, colder, more or less snow, more or less wind, greater or lesser amounts of honey or pollen on hand as winter begins, greater or lesser population — these and other factors vary from year to year and influence the condition and strength of the colony. For instance, a relatively warm winter can mean higher consumption of stores because of increased activity in the hive. A relatively colder winter can mean higher consumption too, simply to sustain life in the harsher temperatures. Each variation in conditions can have its subtle effect. The beekeeper must remain aware of all of the conditions that may affect the bees and should be ready to take some concrete steps.

It is easy to say that each winter is different, but difficult to remember specifically how this winter differs from the last. One of the best ways to relate to the past is to keep a beekeeping diary. This can be as simple as periodic notations on a calendar, or as detailed as daily entries in a journal. Precise records on nectar flows, weather, requeening, amounts and timing of feeding, and the onset of specific problems can be invaluable. Memory is fallible. Give it a boost.

Winterkill

First, on this beautiful day in February, determine if any colonies have died. A colony may actually be dead, but robber bees from another hive flying in and out give the illusion of well-being. Open the cover and be sure that there is a cluster of resident bees. If the colony is in fact dead, brush out the loose, dead bees from between the combs and from the bottom board. Left in the hive, these would rot and attract scavenger insects and mold as the weather warms up. Seal the entrance or remove the hive to a bee-tight area where it is impossible for bees to rob any honey that may be in the combs. This precaution is important to prevent the possible spread of American foulbrood or any other disease or problems that may have caused the demise of the colony.

Be absolutely sure that you know why your colony died. If it was because of disease, and especially American foulbrood, you must know and must take appropriate steps before attempting to reestablish a colony in that hive. If you cannot tell why your hive died, get an expert opinion.

There are three basic causes of winterkill: starvation, disease, and queenlessness. If a colony has starved, the cause is fairly obvious. There will be no food in the hive, or perhaps there will be some food but it is not near the cluster. In severe weather it is possible for the cluster to become isolated from the food stores, either because the cluster contracted in response to the cold, or because it ate all of the food within immediate reach and, due to the cold, was unable to move on to adjacent stores. With starvation there are usually many dead bees still in the colony, in the cluster or on the bottom board. Whatever food is in the hive is shared equally right up until the end so that the bees die off more or less together. This is not usually true of the other causes.

Disease and queenlessness usually result in a dwindling of the population over a period of time. With queenlessness, no new bees are added to the colony and natural die-off slowly reduces the population. With disease, depending on the specific ailment, brood rearing may cease or new bees may be born but at a lower than normal rate. With either queenlessness or disease, the actual cause of death will probably be freezing. Bees keep warm in the winter by clustering, eating honey for fuel, and shivering their flight muscles to generate heat. However, there is a minimum size for the cluster below which they are no longer able to generate enough heat for survival. This minimum size is variable, depending on the surrounding temperature. As the number of bees dwindles, the cluster loses the capability to keep itself warm. The bees will simply die of the cold.

Assuming that your colonies are alive and well, clean the entrance of each hive of any accumulation of dead bees or other debris. Thousands of bees die naturally over the winter. The workers will clean these out if they can, but, depending on the weather and the presence of mouse guards or other restrictions, varying numbers of dead bees accumulate on the bottom board, sometimes blocking the entrance. If you clear the entrance periodically you ensure continuing passage for the bees on their cleansing flights, and allow air movement for hive ventilation.

Feeding

Inspect each hive for stores. With the peak of brood rearing approaching, there should be several frames of honey and pollen on hand, and these should be immediately adjacent to the cluster so the bees can get at them. If there is intervening space, move the stores closer. Do not, however, disturb the cluster.

If stores are low, feed. Frames of honey saved in the freezer from the previous fall are best. Kept in the freezer, the honey will not granulate and will be most readily used by the bees. Be sure that the frames are well thawed before placing them in the hive. Bring the honey to at least room temperature. It may not be practical, however, for you to save honey in the freezer. In that case, simply store the frames of honey in a secure place away from insects and rodents and don't worry if it granulates. The bees can still use it, though with a little more difficulty.

If no honey has been saved, then feed sugar syrup, in combs. If you don't have empty combs stored away, simply borrow some from the hive you are going to feed. Take these inside and use a squirt or spray bottle to fill the cells with syrup. Simply pouring syrup on a frame is messy and inefficient. Much of the syrup runs off and is lost. Squirting it under modest pressure is better. When you're dealing with just a few frames a hand-operated squirter, such as a liquid dishwashing soap container, does a fine job. The syrup can be forced into each cell. Use warm syrup and if possible place it in the hives while it is still warm.

During the winter months the bees will form a cluster, within which, except for a period in early winter, the life of the colony will proceed with some normalcy. In the center, brood rearing may be underway. In February, as we make this initial inspection, it almost certainly will be.

When feeding frames of honey or syrup, put them as close to the cluster as possible. Again, do not disturb the cluster, and be especially careful not to disturb the integrity of the brood nest. The size and shape of the brood nest and the cluster are such that the bees can keep everything warm and allow brood rearing to go on even with severe temperatures outside.

If the bees are flying regularly, syrup may be fed in a top feeder or division board feeder, or in one of the other types of internal feeders that beekeepers have devised over the years to place the feed close to the bees. However, it is not advisable to use a Boardman feeder at this time of year. In addition to being outside in the cold, a Boardman feeder is far removed from the cluster in a normal hive. It is important that the feed be kept as close to the cluster as possible so that the bees can get at it even during a cold snap or at night.

In cold weather sugar candy may also be fed, and in a desperate situation, dry granulated sugar may be used. Both of these are more difficult for the bees to eat than are honey or syrup and are not highly recommended. The hard sugar candy is normally made in thin sheets and may be laid directly over the bees, either on the frame tops or on the inner cover. Dry sugar is usually fed on the inner cover or in some type of trough arrangement, again, as close to the cluster as possible. A recipe for mixing sugar candy may be found in Chapter 16. Note that it does not include cream of tartar or tartaric acid, as was recommended in some of the older books. There is some evidence that tartaric acid shortens the lives of bees.

Bees and Cold Weather

All of the foregoing has been postulated on the thought that there will be that beautiful warm day in February and that you can open the hive without fear of harming the colony. But maybe that day doesn't come this year, or it does not come on a day that you can work the bees. Then just go ahead and pick a day that you can work the bees and go to it. It is good to have a nice warm day but it is by no means imperative. Bees are a lot tougher than we give them credit for, and the colony can take more abuse than most of us realize. Just as a person can tolerate being outdoors briefly in adverse weather, so can a colony of bees tolerate being opened briefly under similar circumstances. Temperatures in the thirties and forties are not too cold to open, inspect, and feed hives if it is done quickly and judiciously.

Keep in mind, however, that though the weather may in theory be too cold for the bees to fly, some of them can and will fly, though perhaps only briefly. Temperatures in the cluster range from around 45°F at the surface up into the nineties in the center. At least some of the bees are warm enough to fly and will do so until their bodies chill. During that time they are quite capable of stinging.

A further complication is that cold weather makes the propolis seal around the cover especially effective. Opening a well-sealed hive can be a jolting experience for both the bees and the beekeeper when the brittle seal breaks apart. Almost certainly the bees will be alarmed and some of them will take to the air. Depending on just how cold it is, many of the bees that leave the hive will never return, even though they do not sting. They chill, lose the ability to fly, and simply fall to the ground, speckling the snow with their bodies.

Those of us who live in snow country have often seen dead bees on the snow — even on those days when the hive has not been disturbed. It is a startling sight the first time or two that we see it, but it is normal. Bees do die off over the winter, some in the hive, some outside on cleansing flights. To the extent possible the colony removes the dead bees from the hive, dropping them outside, adding their bodies to the ones that flew out never to return. Those bees remaining are the younger generations, all ready to greet the spring as it approaches.

CHAPTER 2

SPRING MANAGEMENT

P ERHAPS THE MOST IMPORTANT ADVICE TO BE GIVEN regarding spring management is this: do not rush the season. After that long cold winter, on one of those inevitable warm, sunny days in early April when you just know that spring is finally here, it is difficult to resist that urge to get out there and do something with the bees. But think back to other years. Is spring really here? In some years it is, but in many, many more it is not. There are still some cool, rainy, perhaps unpleasant days yet to come. In some areas it is quite possible to have a little snow in April, and little if anything is in bloom. So — except for feeding — do not rush the season. A good signal to look for is dandelions. When they start to bloom, it is time to begin your spring chores.

What are these spring chores? First, remove mouse guards and any winter packing or insulation. Then check colonies for food, disease, and the quality of the queen. Feed, if necessary. In mid to late spring, the time of heaviest brood rearing, prodigious quantities of stores are used, both pollen and honey. A significant number of colonies are not able to raise brood to their full potential at this time for lack of food, and some colonies die of starvation. The methods for feeding that were discussed in Chapter 1 are still valid, but with warmer weather here you will most likely feed syrup in a top feeder. It is still too cold in the more northern areas for effective use of a Boardman or entrance feeder. Once you start feeding, be sure ample food is on hand at all times. A break in the availability of food can mean a break in the brood cycle. The colony will not raise brood if it does not have food reserves, and if food runs low or runs out they may remove all or part of any brood in the hive.

Inspect the brood carefully. Does it appear healthy? Is there a proper balance of eggs, larvae, and pupae? Is the brood pattern full and uniform? The answer to each of these questions must be weighed against a variety of factors. For instance, if the brood does not appear healthy is it because of disease for which you must treat, is it a failing queen, or is it perhaps chilled brood that died in a recent cold snap? Chilled brood of course is not a disease. It results from a sudden change in the weather and an overextended brood nest, one too large for the colony to cover in unseasonable cold.

The brood patterns should at all times show a seasonal balance of the three stages of brood — eggs, larvae, and pupae. In the spring, for instance, with population increasing, egg laying should be at an increasing rate. As with so many things in beekeeping, it takes experience to recognize that a proper balance exists. If you don't have that experience, simply study the brood area carefully and store away your impressions to be weighed against the success of the colony and against similar observations in the future. Look for the following:

■ Bees occupying from 12 to 15 frames.

■ Brood in 6 to 10 frames. They won't all be full. Remember, the brood nest is spherical.

■ At least 20 lbs of honey — the equivalent of four deep frames full.

■ At least 2 to 3 frames with pollen.

If you have anything substantially less than this your colony can be considered under strength. If the colony is not strong now it may not have the underlying vitality to build up as the season progresses. It is time for you to intervene. The specific action you take will depend on two things: your resources, and the condition you are correcting.

Let's consider resources, that is, sources of additional bees. If you have only one colony, your personal resources are limited. Perhaps, though, you have a beekeeping buddy who is willing to share, or there is a ready source in your area where you can purchase queens and bees. Some of your possible actions are:

■ Add a frame or more of capped brood. Capped brood requires minimal care from the receiving colony, so it will not be a drain on its presumably limited resources. This brood will also emerge sooner than will uncapped, contributing working members to the colony sooner.

■ Requeen with a young, well-bred queen from a proven source. A new, young queen will give a burst of new life to the colony — provided the colony has the resources to support that activity. The amount of eggs that a queen will lay and the amount of brood that a colony will raise are a function of the food and population resources of that colony. Surplus eggs or brood will be destroyed or eaten. You may wish to add a frame of capped brood along with the new queen to ensure that there is sufficient population to support her.

■ Add a small package, with or without a queen. This is a possibility in the event that brood is not available to transfer from another colony.

■ Combine with another colony. If it is not possible to add more brood or bees to a particularly weak colony, it may be best to combine it with another colony. Some colonies are simply not strong enough ever to build up to a satisfactory level. Such a colony is a prime target for disease, wax moths, or robbing. It is better to save the bees, the comb, and any stores that may be present than it is to let the colony dwindle and die. This is an action that many beekeepers, novice and experienced alike, have great difficulty in accepting but in the long run it is best. Done early enough, the emptied equipment is then ready to house a swarm if one presents itself.

Feeding of course continues as long as necessary. Assuming there are no honey supers in place, there is no harm in feeding even if the colony has food reserves. The syrup will act as a stimulant to colony activities, especially if it is fed in a Boardman type of feeder, or any other that allows the syrup to trickle in, simulating a nectar flow. If the syrup is surplus to the colony's immediate needs, it will be stored and used later, allowing more honey to be stored for the beekeeper.

Reversing

Reversing is a normal part of spring management. It is a practice that is often misunderstood, however, and consequently often misapplied. Proper and timely reversing of hive bodies can be of great benefit but done too early, too late, or too often, reversing can be a setback for a colony.

Reversing is the simple act of switching the two hive bodies of a colony, one with the other. It is normally done at about the time of the first significant nectar flow of the spring. It is normally done only once per season. It is done only if the brood is at that moment occupying only a single hive body. Let's take these points in order.

Timing

In the early to mid-spring the typical colony will be found occupying the upper hive body in a two-story hive. The lower body will be empty or nearly so. The natural utilization of a hive is upward. Over the winter the colony has worked its way from the bottom to the top. Its inclination is to continue moving up. If it cannot, it will begin to feel constrained, and this can contribute to the onset of the swarming urge. Reversing should take place then before the swarming urge begins to develop. However, it must not be too early. Spring weather is variable. Though the bees may be active and there may be nectar available, there are cold days still mixed in and certainly cold nights. Brood rearing is well underway and the brood nest is expanding. Reversing too early places a large volume of empty space above the cluster and brood nest. The bees must work harder to keep the cluster warm because heat is continually being lost into that as yet unused upper hive body. Too late and the swarm urge may begin to develop. The beginning of the dandelion flow is a nicely balanced time. The colony is about ready for the extra space above but the pressures of an approaching swarm season are still light.

Frequency

In a normal colony with a vigorous queen, after reversing, the brood nest will expand slowly but steadily into the upper hive body. The bees will also store honey and pollen in that upper body. It will quickly become fully utilized and the colony will be properly organized with the brood nest, honey, and pollen each in its proper place. And each has its proper place, with the brood nest centralized and towards the bottom, a shell of pollen to the top and sides, and a thicker shell of honey to the top and sides of the pollen. This has evolved over millennia as the most efficient method of hive organization and the bees will go to great lengths to maintain this arrangement.

One school of thought advocates reversing hive bodies every two weeks or so throughout at least the first half of the season. I have seen it recommended in writing more than once, but unfortunately there has been no qualifying explanation with any of these recommendations. I am sure that this has resulted in some beekeepers going to their hives every two weeks and reversing the hive bodies with no thought or knowledge of what may be happening inside that hive. This practice is almost guaranteed to set the hive back.

Occupancy

Assuming that the hive was reversed at about the time of the first significant nectar flow, two weeks is plenty of time for a normal colony to make substantial use of the second hive body. It is unlikely to be completely full but there will be enough progress so that the normal pattern of honey stored over pollen stored over brood is well established in the two hive bodies. Indiscriminate reversing can do two things. It can upset the normal use pattern in the hive, which the bees may spend valuable time correcting. It can also break up the integrity of the brood nest.

Brood nest integrity is vital. At any time, the size of the brood nest is a function of the size of the adult population. The colony will raise no more brood at any given time than they can feed and keep warm. The normal brood nest is more or less spherical. This is the

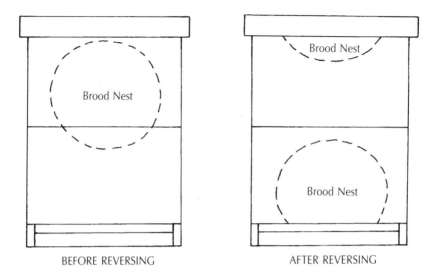

| BEFORE REVERSING | AFTER REVERSING |

FIGURE 2-1. Here the brood nest extends into two hive bodies. If the two bodies are reversed the integrity of the brood nest is destroyed. It is more difficult for the bees to tend it and keep it warm. In unseasonably cold weather brood may be lost.

shape that gives the least surface area for the mass involved and is therefore the most efficient shape possible to keep warm. Breaking up the brood nest as can happen with indiscriminate reversing will probably destroy this efficiency. Coupled with unexpected cold weather, there is an excellent chance of losing brood.

Once the initial reversing has been done, the aware beekeeper should monitor the utilization of the hive. If excess empty space is developing in the lower hive body, judicious shifting of individual frames is probably the correct move to make. Taking care to maintain the basic integrity of the brood nest, brood can be consolidated in the bottom and honey in the top.

As the season moves along, continue monitoring. But recognize that if a significant amount of empty space is found in the food or brood chambers on a continuing basis, there is probably a basic problem with that hive that reversing will not correct. The colony may have a poor or failing queen, be inherently weak, be diseased, or simply never have recovered from a difficult winter. It also might reflect normal conditions for your particular area, or it may be that you are experiencing a poor nectar season. Consult your diary (you are keeping one, aren't you?) to see what the comparable situation was in previous years. Most importantly, do something. Do not just sit back and hope for the best. The actions that you might undertake are those outlined earlier in this chapter: feeding, requeening, adding brood, or otherwise strengthening the colony.

Comb Replacement

Over the years I have occasionally encountered a beekeeper who was very proud of the well used, black comb in his hive. One such bee-keeper bragged of his comb, saying it had been his grandfather's, and he intended to use it for years to come. He didn't say that he planned to pass it on to his grandchildren but I suspect that idea was in the back of his mind. In recent years, however, studies have shown that old, black comb is a likely source of problems, primarily disease organisms, and should be replaced, usually with new foundation. This being true, we have the question — how often should comb be replaced? How black can we allow it to become? There is no pat answer to this question. There are some thoughts and guidelines though.

The blackness of brood comb comes from a number of sources — propolis, cocoons, larval feces, and the dirt and travel stain resulting from constant use. There also may be disease organisms or pesticide residues mixed in, sometimes harmless, sometimes not. Some of these organisms and residues are sealed into the comb and will probably never cause any problems. Others are not yet present in sufficient numbers to be a problem. But as the comb grows older the chances become greater that the numbers will build up to a point where they are potentially dangerous. Remember, disease is probably present in every hive. The apparently healthy colonies simply don't have enough of these organisms to create an infection, or the colonies are disease resistant.

Aside from the potential for disease, there is another drawback to old comb. Each time that a cell is used for brood rearing, a cocoon is left in the cell, adhering to the cell walls. The bees are not able to remove this cocoon. Between layers of cocoons are larval feces, compounding the buildup. Over time, as comb is used for brood rearing, the individual cells slowly become smaller. As a result, smaller bees are raised in those cells.

By replacing the comb periodically the beekeeper not only makes the hive a healthier place to live, but also allows for larger and presumably stronger, more capable workers, who gather more nectar and pollen.

Further, studies have shown that queens prefer new comb for laying eggs, and workers prefer darker comb for storing honey. Replacing old dark brood comb can thus help in hive management in two ways: it encourages the queen to stay in the brood area, and encourages the workers to store honey outside of the brood area, where initially at least the comb is a little darker. This in turn should help to reduce swarming by keeping the various hive activities in their proper places and minimizing the sense of crowding.

This gets us back to the original question: how often should comb be replaced? One approach is to replace two of the oldest combs from each hive body each year, perhaps more than two the first couple of years to get the program underway. If you replace two per year, then you are on a five year cycle. This may or may not be right for you. Some beekeepers may prefer a three or four year cycle. Again, individual situations must be considered. If your hive is relatively new and no comb is really in need of replacement, you might go ahead and replace a couple anyhow, setting aside the culled comb for possible future use. It's always handy to have some spare drawn comb available.

Finally, when is the best time to replace comb? Do it in the spring, when the hive population is relatively low and many combs are still empty from the long winter just past. Anticipate this replacement, however. As you work your hive during the summer, mark those frames that are candidates for replacement and shift frames as the opportunity occurs so that those to be replaced are on the outside edges of the hive. This reduces the possibility that the frames will be in use when replacement time comes the following spring. Perhaps you can work it so that combs to be replaced are all in the bottom hive body in the fall. The bees will normally have worked their way into the upper hive body during the winter, leaving all of the frames in the bottom empty.

A final note: if you are planning to pass your hives on to your grandchildren, they will appreciate it if the combs have been kept culled.

Population Cycle

The population of a colony of bees is not constant throughout the year. In the active season, when the bees are needed, the population is up. In the winter, when they are not needed, the population is down. This is not surprising. It is good business. In the summer, when nectar and pollen are available, a large population is most desirable. In the winter, when hive activities are at a low level and food supplies are limited, a smaller population is best.

The beekeeper needs to know what the relative population of a colony should be at any given time of the year. This is one of the factors used in evaluating the condition and status of the colony. Population is a function of brood rearing, of course, so the beekeeper should also know the levels of brood rearing appropriate for a given season of the year.

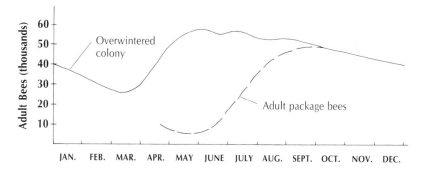

FIGURE 2-2. Colony Population Cycle

Figure 2-2 shows two curves, one representing the population cycle of an overwintered colony and the other the population cycle for a package installed in the spring.

The cycles for swarms and nucs could also be shown in a similar manner, but the size and the timing of hiving or installation is so variable that no one curve would suffice for either. Note that the curve for the package dips initially, just as it would do for a swarm. This is because packages and swarms normally have no brood when first installed. The first eggs laid will not mature and hatch for at least three weeks. During this period, some of the original workers are reaching the end of their normal life cycle and they are dying. The population will decline until new bees begin emerging at a rate that exceeds normal die-off, four or more weeks from installation.

What is the practical value of knowing about population levels and cycles? Population is a strong indicator of the present and future health and well-being of a colony. For instance, a colony that is under strength in June is going to have a difficult time storing enough honey for winter. Population normally peaks in June and that peak population puts away the beekeeper's honey as well as its own winter stores through the balance of the season. On the other hand, a colony that is exceptionally strong in the late season may also have a problem. It may not be able to put away enough winter stores for such a large population, simply for lack of space. The aware beekeeper can correct or compensate for both of these situations. We will address these and other problems at appropriate places throughout the book.

CHAPTER 3

SWARMING

∎

SWARMING, A NATURAL PHENOMENON of a honey bee colony, is greeted with a wide range of reactions by beekeepers. For some, swarming is inevitable, and there is no way to stop it. For others, it is seen as a measure of beekeeping competence, in themselves or in others. A competent beekeeper would never allow swarming, so they think. For still others it is a welcome event, assuming that the swarm is captured. It gives the beekeeper the opportunity to start a new colony. It is especially welcome, of course, if the captured swarm did not come from the beekeeper's own hive.

Reasons for Swarming

Aside from any questions of desirability, competence, or attitude, what is swarming? It is a natural part of the yearly cycle of a honey bee colony, developed by nature to perpetuate the species, to compensate for losses of colonies by natural causes such as forest fire, disease, starvation, and the like. Not every colony swarms every year. Left to its own devices the average colony would probably swarm once every two or three years. Of course, kept bees are not left to their own devices. Keeping bees in a hive puts unnatural pressures on the colony and the beekeeper should have the possibility of swarming in the back of his or her mind at all times. And that does not mean just during swarm season. A beekeeper does have a level of control over swarming, but it requires that elements of this control be practiced on a continuing basis over the beekeeping year.

The swarming season occurs during a more or less set period each year. In southern New England, for instance, the primary swarm season runs from around mid-May until the end of June. About 80 percent of the swarms for that year will emerge then. There is also a secondary swarm season, less well known to many beekeepers who express great surprise when a swarm emerges then. This secondary season runs from approximately mid-August until mid-September, and about 20 percent of all swarming occurs then. Swarms from this secondary period are far less likely to succeed in their new homes. Rarely are they able to establish themselves and build stores and population sufficient to ensure a successful first winter. The parent colony may also have difficulties, though they at least have an established home and, presumably, food stores on hand.

Moving north or south, the dates of the swarming season change, becoming later to the north and earlier to the south. Moving north also causes the swarming urge to intensify. Presumably this relates to the increased amount of winterkill in the harsher climates of the north. With more colonies dying, more swarms are desirable as replacements in nature.

Causes of Swarming

So far we have been talking about the reason for swarming — perpetuation of the species. A different matter is the cause of a given swarm. The causes of swarming are not completely understood. There seem to be several interrelated factors, and the exact nature of these relationships is not clear. The factors include colony size, congestion, worker age distribution, and the queen. Because we do not completely understand the relationships between these factors, swarm control continues to be a chancy thing. Let us consider the factors individually.

The Queen

Activities over a period of weeks, sometimes months, have a bearing on whether or not a particular colony will swarm. However, the bees' actual "commitment" to swarm is made when they commence rearing queens. Normally, queen rearing is inhibited by the presence of a certain level of queen substance in the hive. This substance is secreted by the queen, picked up by her attendants as they groom and feed her, and distributed through the hive by the continuing interactions of the workers as they go about their daily routine.

Various conditions, either actual or perceived, can bring about a reduction of this level. A congested brood area with few empty cells slows the queen's movements so that her contact with workers and the dispersal of queen substance is slowed. A crowded brood area where workers are overabundant and idle slows down the transfer of queen substance from bee to bee. An aging queen will secrete a lesser level of queen substance. These conditions can be related to colony size, congestion, and worker age. Control and prevention of swarming, which we will consider shortly, must recognize all of these factors.

Colony Size

Many beekeepers assume that only colonies with large populations will swarm, but this is untrue. I have seen a four-frame nuc swarm, and have heard of a small observation hive swarming. In fact, swarming usually takes place before a colony has reached its peak population for the season.

In the context of swarming, colony size really has three components: the volume of space available, the amount of comb in that space, and the bee population. An imbalance of these three components contributes to the onset of swarming.

Congestion

Even with large areas of unused space in a hive, if the brood area is congested we have taken a long step towards swarming. This congestion normally occurs in two ways: first, when all or most of the available brood cells are used, and second, when large numbers of young bees are hanging about the brood nest, and, because of their numbers, have little to do.

Worker Age

Swarming season occurs at a time when population is rising, usually rapidly. Many young bees are emerging, so that they become a large proportion of the total population. Young bees, whether working or idle, contribute to congestion by centering their activities in the brood area.

The Act of Swarming

Preparation

The process of swarming begins well before the actual event. Signs may be visible to an alert beekeeper weeks before the fact. Be particularly observant in the early spring when you may find some combination or perhaps all of the following conditions:

■ the winter has been easy;

■ spring is early;

■ plentiful food reserves are on hand;

■ the queen is old;

■ drone rearing is early.

Later, as population continues to build in the spring, watch for queen cups. Queen cups, as opposed to queen cells, are almost always present in a hive. Their presence alone does not mean that a colony is going to swarm. However, they do add to the possibility when the other signs are present. The more queen cups present, the greater the possibility.

Just prior to the actual event, in the few days before the swarm actually leaves, the queen will taper off her rate of egg laying. About three days before the departure she will stop laying altogether. During this period she will be getting down to flying trim, losing as much as one third of her normal weight. Workers, meanwhile, will do less foraging and more hanging around, especially at the hive entrance.

Immediately prior to departure there will be much unrest and little foraging, and the workers will engorge on honey. Though not all the workers will leave with the swarm, the preparatory activities involve the whole population. It is during this period that scouts begin searching for possible nest sites for the swarm. Normally, when the swarm actually departs the bees are aware of at least one potential new home. The final selection usually takes place after the swarm has left the parent colony and has clustered near the parent hive.

Since the queen mother departs with the swarm, steps must be taken by the parent colony to see that it has a new queen. Accordingly, seven or eight days prior to the time that the swarm ultimately leaves, the colony will begin raising several new queens. Weather allowing, the swarm departs on or shortly after the day that these queen cells are capped, leaving the parent colony without an adult queen for about a week. Consider this any time that you are about to cut capped queen cells during swarm season. Not only is it probable that no queen is present, but under normal circumstances there are no eggs and, possibly, no young larvae. Remember, the queen stopped laying a few days before the swarm actually departed. Cutting queen cells now may leave the colony hopelessly queenless.

Departure

On the big day, usually around midday, the momentous event occurs. There is increasing agitation around the entrance and then finally the bees come pouring out. Large numbers of bees participate in the initial rush, though not all of them will necessarily join the swarm when it actually leaves. The queen will emerge somewhere in the middle of the pack: she is a follower, not a leader. After swirling about for several minutes in the general vicinity of the parent hive, the swarm will settle and form a cluster on some nearby protrusion, for instance, on the branch of a tree or shrub. At this point the swarm determines if the queen is in fact with them. If she is absent they will return to the parent hive, perhaps repeating the departure process one or several times until they are joined by the queen or until they give up trying.

Assuming that the queen is present, the swarm will next come to agreement on the location of their new home. Scouts have been out, probably for several days, and have usually located one if not several potential new homes for the swarm. These scouts now begin dancing on the surface of the cluster, each extolling the virtues of her selected

EXPOSED COLONY. These bees swarmed but apparently could not find a suitable cavity. They have established permanent residence in the midst of a shrub. They are pictured three weeks after swarming.

new home site. At some point the entire swarm comes to agreement as to which of the new sites to accept and the swarm departs to take up residence.

It is during this period in the life of the swarm that the beekeeper is able to pick up the swarm and establish it as a new colony. The period may last from a few minutes to two or three days. Rarely, the swarm will never move on, apparently unable to locate a suitable cavity in which to take up residence. They will begin to build comb and establish a nest wherever they may be, usually hanging from a tree limb or in some other exposed situation. Such a swarm rarely survives past the first winter, especially in the colder parts of the country, since it has no protection against the elements.

Swarm Control and Prevention

The Swarm Impulse

Conditions all being right, a colony of bees that is going to swarm will develop a swarm impulse. That is, at some time before the actual swarm season begins, they seem to develop a mind-set towards swarming. The closer they get to the actual swarm season, the stronger the impulse becomes. The beekeeper's task is to prevent that impulse from developing. Up to a point, this can be done by correcting conditions in the hive — a new queen, more space, fewer bees, and other manipulations. These are not last-minute actions though. They must be done early so that the bees do not get ahead of the beekeeper and swarming becomes inevitable. If this does happen, then the beekeeper's recourse is to satisfy the swarm impulse artificially. Differently stated, the beekeeper must make the bees believe that they have swarmed.

When a colony does swarm we then have two colonies of a fairly definite makeup of bees. The parent colony is left at the old stand with all of the existing brood, many field bees, a lesser number of house bees, and no queen, though one will emerge in a few days if nothing goes wrong. The new colony, once it moves into its new quarters, will have a mature laying queen, a large number of young bees, a more limited number of older bees, and no brood.

Beekeepers use many practices to keep their bees from swarming. Some of these are successful, some are not. Some of them are successful part of the time but not always. A key here is that beekeepers too often do not recognize the existence of the swarm impulse, or do not recognize the need to work with it. They may do the right thing but at the wrong time, usually too late. Again, once the impulse is there, it must be satisfied. Adding a super or removing a little brood is not effective at the last minute. To make this a little more clear we need to define some terms.

Two terms frequently used relative to swarming are swarm control and swarm prevention. Though often used interchangeably, they do not mean the same thing. For our purposes we will use the following definitions:

■ Swarm control — the ongoing management practices used throughout the year to keep a colony from wanting to swarm, from developing the swarm impulse.

■ Swarm prevention — the last minute practices that must be implemented to stop swarming after the swarm impulse has set in. The need to practice swarm prevention means that swarm control did not work.

Swarm Control

Swarm control measures tend to be positive actions. That is, they are actions that are in keeping with good hive management. Some of the normal swarm control practices that a beekeeper might follow are:

■ Requeen regularly. A young queen is secreting a high level of queen substance.

■ Maintain good comb. Cull excess drone comb. When in the hive, drones normally congregate in and around the brood nest. Drones are necessary to the well-being of any colony but an excess of drones contributes to congestion in the brood area.

■ Reverse hive bodies in the spring. This helps to relieve congestion by giving the bees space above the brood nest into which to move.

■ Do not allow the colony to become honey-bound. Smaller colonies tend to store honey close in to the brood area, creating a barrier that they cannot or will not break through. This barrier will restrict and congest the brood area. Shift frames of honey as necessary to relieve the pressure.

■ Maintain strong colonies. A strong colony usually expands readily and distributes itself evenly through the supers. A strong colony will push through a honey barrier if one develops. Maintaining strong colonies, of course, is what this book is all about. Such actions as requeening, timely hive manipulations, disease control, and other activities discussed later are all steps to maintaining strong colonies.

■ Super up in plenty of time. Keep supering ahead. The tendency to swarm is strongest early in a honey flow. If the bees are into the supers early, congestion is less likely to develop.

Other factors that may have a bearing on crowding and congestion, and therefore on swarming, relate to hive location and exposure. Sun, shade, and ventilation can all affect a colony's well-being. Continual full sun, deep shade, or poor ventilation can all have effects that increase the sense of crowding in the hive.

Swarm Prevention

Swarm prevention actions are intrusions into colony life. None of them are in keeping with maximum brood or honey production. First we will deal with some negative beekeeper actions. With these, the colony is almost certain to keep trying to swarm until it is finally successful, or it will eventually give up trying but will be so weakened that it might have been better off had it been allowed to swarm and rebuild itself. In this category we have:

■ Removing or caging the queen. Every day that the queen is out of circulation means a loss of 1,000 to 1,500 eggs to the colony. As the days pass the amount of brood in the hive slowly diminishes. At some point the swarm urge may in fact be satisfied by virtue of the lost brood but the overall population will be much reduced.

■ Queen excluder to confine the queen. An excluder under the bottom hive body has the advantage of leaving the queen free to lay eggs but it is otherwise inhibiting. In many colonies, the bees work much less enthusiastically through an excluder. Further, drones cannot pass through it. As time passes they will clog it up with their bodies, both live and dead, contributing a form of congestion. Meanwhile, population will probably continue to build and the pressure to swarm will be maintained. At some point, for the good of the colony, the excluder must come off. Since the swarm impulse was never satisfied, a swarm very likely may leave.

■ Destroying queen cells. To be effective, the beekeeper must open the hive, disrupting it totally, every five days. Because of the timing of queen rearing and of the act of swarming, anything less often will not be effective in stopping a swarm. In searching for queen cells, every one must be found, and in a populous hive this can be difficult. Allowing one cell to slip by uncut is enough to release that swarm. Of course the constant turmoil of the beekeeper's inspections is very disruptive of production. Again, it might be better to let the colony swarm and be done with it.

■ Returning a captured swarm to its parent colony immediately. If this is done the swarming urge will not have been satisfied and they almost certainly will swarm out again.

The second category of swarm prevention actions, which we are terming positive, are still not in the total best interest of the parent colony, though in the long run they are usually in the beekeeper's best interest. Each of these actions involves removing brood or adult bees from the parent colony in such a way as to make the colony as a whole believe that it has swarmed. Some have a permanent effect on the parent colony. Others are temporary.

Permanent change. The result of each of the following actions is basically the same as if a swarm had taken place. The colony remaining on the original stand may be configured as a parent colony, or as a swarm colony in its new home. Either way, the swarm urge has been satisfied. These actions include:

■ Making a split or a nucleus hive. (See Chapter 5, Making Increase)

■ Removing frames of brood and attendant house bees, which may be added to a weak colony.

■ Swapping a weak hive with a stronger one that is on the verge of swarming; that is, exchanging the positions of two hives one with the other. The weak hive thereby acquires the field force of the stronger hive, helping it to build up. The strong hive is now unlikely to swarm but it still has its larger house bee and brood population and should quickly rebuild.

Temporary change. At some point in every beekeeper's life there comes that moment when he opens a hive and realizes the horrible truth: swarm control has failed. This hive is going to swarm at any moment. There are capped queen cells present. There are few or no eggs. The bees are restless and hanging about. Little foraging is in progress. Do not despair; there still is hope.

To compound the problem, the beekeeper may have only the one colony and does not want any more. There is *still* hope. Over the years beekeepers have devised a number of manipulations that may be used to make the bees believe that they have swarmed while maintaining the basic integrity of the parent colony. I will describe one of them. You may devise any number of variations.

Two extra pieces of equipment are required, a queen excluder and a double screen or Snelgrove board. Reconfigure the hive as shown in the illustration.

You now have in effect two colonies, one on top of the other.

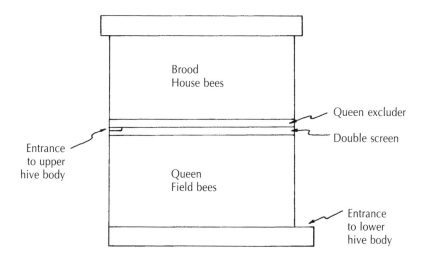

FIGURE 3-1: A colony on the verge of swarming is split temporarily. The two halves of the colony are now semi-independent and each half thinks that the parent colony has swarmed.

SNELLGROVE BOARD. **This double screen board has the lower of its two rear ports open.**

The top hive body contains the house bees and all or most of the brood. The lower hive body contains the queen, the field bees and perhaps a few of the house bees. There may be some capped brood if it all did not fit in the upper hive body. Check carefully here and eliminate any capped queen cells.

Entrances for the two colonies are as shown. The lower colony continues to use the original entrance, thereby keeping all of the field bees. The upper colony has a new entrance at the rear, utilizing one of the ports in the upper side of the screen board. For insurance, a queen excluder is placed over the screen board. In theory the upper colony will destroy any queen cells overlooked by the beekeeper and will not swarm since it has no field bees. It is best, though, to be sure.

Supers may be placed over either or both colonies as needed. After about 5 to 7 days, remove the screen board and the queen excluder and restore the hive to a normal configuration. The swarm urge will have been satisfied and life will continue. Give some thought to why the colony was preparing to swarm. Perhaps it was because of a failing queen, in which case it is advisable to requeen now.

Capturing Swarms

There will always be swarms — if not from your hive, then from the hives of others, or from a feral colony in the area. Consider a swarm a resource. Do not ignore it or give it away as many beekeepers do. Capture it if possible and use it to start a new colony or to strengthen a weak colony. If for no other reason, capture it to keep it from terrorizing the neighborhood. In general the public does not understand swarms and you can become a hero.

In an emergency a swarm can be picked up in almost any kind of a container of suitable size, a cardboard box, a pail, whatever is handy. A means of closing it tightly is important, but not so tightly as to suffocate the bees. There must be ventilation. It is also helpful to place a stick diagonally in the container for the swarm to hang on. Otherwise they are in an uncomfortable mass on the bottom of the container.

This is all fine for an emergency, but in swarm season a little more preparation is in order. Keep some spare equipment on hand — a hive body, even one that is old and tired that you might not normally use on a hive, a bottom board, and a cover. These latter need not be first class equipment either, just something to do a temporary job. For the bottom, a piece of plywood cut to size with strips fastened around three sides will do fine. The top can also be a piece of plywood, with a strip added front and rear to make a migratory style cover. To finish off, use a set of frames that fit the box, with foundation installed. Don't skimp with the frames. These may become part of a permanent hive.

Caution: you may not wish to start a new colony but would rather add your newly captured swarm to an existing colony. Do not do this immediately. Be prepared to set the swarm up on its own, preferably in an isolated yard, where you can monitor it for the presence of disease or mites. This is especially important if you are uncertain of the origin of the swarm.

CHAPTER 4

THE QUEEN

HAVING A PROBLEM OR TWO with your hive? Population a little low, or honey production not what it should be? Perhaps you're having more than your share of swarms, or maybe the colony is a little more aggressive than you would like. These and other shortcomings of your hive could be the fault of your queen.

All queens are not equal. What do you really know about yours? There are several things that you should know, either positively or with some degree of confidence. For instance:

■ Her age. Is she young and vigorous, laying an optimum number of eggs per day, or is she older, slowing down, and ready for supersedure?

■ Her origin and lineage. Was she identified as a potential queen at a sufficiently young age (egg or young larva) or was she past her prime (an older larva)? Is she the offspring of a quality queen mother? Was there some control over the drones with whom that queen mother mated?

■ The conditions under which she was bred. Was she raised in a colony with plenty of nurse bees and ample food, in a well formed cell of ample size?

- Mating. What were the conditions under which she was mated? Was there good flying weather when she was ready to mate so that her mating flights were timely? Were there other colonies in the area to provide drones, reducing the possibility that she mated with her own brothers? Did she mate with enough drones so that she is carrying a lifetime supply of sperm in her spermatheca?

- Colony demands. What have been the demands on her as an egg layer? Have they been such that she has been laying at capacity or near capacity regularly, or is this a weaker hive that has not required the queen to live up to her full potential?

The kinds of questions asked here for the most part do not have obvious answers. They require some thought and some understanding of the life of the colony and of individual bees.

Though all three castes of honey bees are important, the queen is usually accepted as the most important bee in the colony. She is unique. Under normal conditions she has no immediate backup. Both her presence and her absence have many implications for the colony. Beekeepers do not always consider these implications, perhaps not even being aware of some of them.

The queen is the mother of all of the other bees in the colony, assuming she has been present in the colony as a laying queen for at least six weeks. This fact does not seem important to many beekeepers, and perhaps it is not. Occasionally, however, there are situations or events where it is significant, and the beekeeper should have some awareness of it. These situations include: correcting a "mean"[*] or aggressive colony, changing the race or strain of bees in a colony, or preventing inbreeding. Let's take these points in order.

A Mean or Aggressive Colony
Meanness or aggressiveness can be caused by outside influences such as hive location, exposure, or harassment by people or animals. The

---

[*]*Describing bees as being mean is anthropomorphic. Bees do not get mean or angry or have other human traits. These are simply convenient handles by which to describe particular behaviors or degrees of behavior. Bees simply react to particular stimuli at varying levels of intensity.*

cause can also be genetic, however, inherited through the queen. Before correcting the problem, therefore, you must determine the cause.

Professional queen breeders raise queens from selected stock, with attention to both the queens and the drones in the lineage. Gentleness is one of the characteristics these breeders work for, since beekeepers normally want gentle colonies. Well-bred queens will usually head up a colony that is gentle. At some point, though, a queen gets old and needs to be replaced. One of several things may then happen. The colony may be the first to detect the problem and take steps to supersede the queen. They will select larvae or eggs laid by that aging queen and raise them, with one of them ultimately becoming their new queen. That new queen will go out to mate and will take potluck from any drones who happen to be abroad at that time. Genetically or temperamentally these drones may be compatible with the queen. Then again, they may not be, and the resulting offspring of one or more of these unions may be mean or have other undesirable traits. This is more likely to happen if the aging queen was herself a hybrid. Hybrids do not breed true and their offspring are more likely to have undesirable characteristics.

Perhaps in the foregoing situation it was the beekeeper who determined that it was time to replace the old queen, and he did so simply by killing the old queen, leaving it to the colony to raise a replacement. Or perhaps he made a split and, again, left it to the bees to raise a new queen. Or perhaps there was not an old queen involved at all, but the queen of the colony, whatever her age, was unknowingly killed by the beekeeper, or died from disease. The colony again is left to raise a new queen, who will mate with whatever drones are available, and again the possibility of aggressive offspring arises.

Other possible situations exist but the point is that there are many ways to bring about uncontrolled replacement of a well-bred queen, resulting in a new queen who mates with drones of unknown or questionable background. Their offspring may be outstanding bees in every way, or they may be something else altogether. This problem is compounded by the probability of the queen mating with as many as ten to twelve drones, giving quite a diversity to the makeup of the colony over time.

Changing the Race or Strain

Many beekeepers, for any of several reasons, may decide to change from one race to another. After getting started with a colony of Italians, perhaps the beekeeper decides to try Carniolans or Buckfast. Or perhaps he hears of a strain that is said to be disease resistant. To change over to the new line is as simple as requeening. The old queen is removed and the new queen from the selected line is carefully introduced. During the active season, workers live for about four to six weeks. Assuming all goes well with the introduction, in six weeks or less all of the workers in the colony will be offspring of that new queen. It does not matter if, for instance, a Buckfast queen is introduced into an Italian colony. The bees themselves do not recognize race or strain. The new queen will be accepted or rejected on her merits as a queen. Once she is accepted, her offspring will be raised as readily as those of the original queen.

Mating With Her Own Brothers

Nature has ways of preventing members of the animal kingdom from mating with their own close relatives. Dispersal is one way: some young animals leave their home territory as they mature and will probably never see their brothers or sisters again. Other animals are born into such a large population that they are not likely to meet a relative by chance later. Others seem to have an inborn mechanism that helps relatives recognize each other and prevents them from mating.

With honey bees, there seems to be more than one mechanism at work. First is a form of dispersal. Mating takes place outside the hive, usually at some distance away, removing the queen from the immediate vicinity of her drone brothers. Second, mating takes place in drone congregation areas. These are areas around the countryside where drones from all the colonies in the neighborhood, including feral colonies, gather and wait for virgin queens to appear. The queens seek out these areas on their mating flights. Some of her brothers no doubt are there but only as a small part of the total drone population. Drones from any one hive do not necessarily all go to the same congregation area, and drones tend to fly farther to mate than do queens, further removing them from each other.

Studies have shown that in colonies headed by queens that did mate with their own brothers, brood mortality is often greater than 50 percent. This mortality takes place at all stages of brood development, with some eggs not hatching, some larvae not pupating, and some pupae never maturing. In addition to the fact that large numbers of potential bees never emerge and add to the population, there is the drain on the colony's resources which went into that non-maturing brood.

If you have more than one hive or your bees are in a neighborhood where there are other hives or feral colonies, this possibility of inbreeding is reduced. If, however, the hive is isolated, you should stay in control by requeening periodically, and perhaps mark the queen so as to be aware if she is replaced by the colony through supersedure or swarming. (See page 37.)

Requeening

There are many suggested methods of requeening, whether the colony is queenless or queenright. No method is guaranteed, but some are without question better than others. One method that is almost certain never to work is to release a new queen directly into the entrance of a queenright colony, expecting her to find and kill the old queen. A method that almost always will work is to introduce a new queen into a nucleus colony that is populated primarily with brood and young bees, and after she is established there, introduce the nuc, frames and all, into the colony to be requeened. To understand why some methods work and others do not, we need to lay some groundwork.

■ Worker bees do not tolerate strange queens, either inside or outside of the colony. Strangers will be at least driven off, but more likely killed.

■ If a beekeeper finds no eggs or open brood in a colony he or she is likely to interpret the situation as queenlessness. If the colony in fact has no queen but is raising one, or has a newly emerged queen that is not yet laying, the bees will consider themselves to be queenright and will not accept a new queen.

■ Workers of a colony may be antagonistic towards the workers

QUEEN MATURATION TIME

Activity	Days from Emergence
Orientation flight	3 - 5
Mating flight	7 - 10
First egg laid	10 - 14

Figure 4-1

inside the cage of a queen being introduced, especially if the introduction is abrupt. That antagonism may be transferred to the new queen if she is released too quickly.

■ In the first few days of her life a newly emerged queen loses weight, becoming less queenlike in appearance, as she prepares for her orientation and mating flights. Just as the worker bees must orient to their hive so they can find their way home, so must the queen. Weather and other circumstances allowing she will make orientation flights as soon as her body has matured so that she can fly, within a few days of emergence. Very quickly thereafter she will make the first of her mating flights and then, finally, will begin to lay eggs. Only then will her abdomen, a queen's most distinguishing feature, begin to reach mature size. During this period the new queen tends to be skittish, running and hiding when disturbed. Because of this and her reduced size she blends in more readily with the workers.

The approximate times involved in a queen's development to maturity are shown in Figure 4-1.

Laying Rate

Another factor that makes requeening go more or less smoothly is the rate of egg laying of each of the two queens involved. A new queen is much more acceptable to a colony if she is laying at about the same rate as the queen she is replacing.

To illustrate how this could affect requeening, let's start with a new queen being raised by a commercial queen breeder. In the queen rearing business a new queen is born in a small nucleus colony with a population of workers that probably weighs less than one pound. The queen makes her orientation and mating flights from this small colony and then starts to lay eggs. This nucleus hive has a small adult population and only a small area of comb available to the queen compared to a regular hive. The queen is left in the nuc only long enough, a few days, to demonstrate that she has mated and is laying fertilized eggs. She is then removed from the nuc, popped into a cage, and shipped. She may be held in the queen cage for a few days at either end of her trip. Her history then when she arrives in the hands of a beekeeper is that she has mated, laid a relatively small number of eggs for a few days, then stopped laying for an indeterminate length of time. She is about to be introduced into a colony.

Using such a queen to replace a fully functional, laying queen in a normal colony will be much more difficult than introducing her to a colony that has no queen or has an older, failing queen. In the former case the new queen, because of her initially lower laying rate, may be perceived by the colony as a failing queen even though she is only two or three weeks old. They may take immediate steps to supersede her. Requeening via an intermediate step, a nuc, will usually eliminate the possibility of this supersedure happening.

To do this intermediate step, the queen should be introduced into a nucleus hive containing at least three frames of drawn comb and young bees. The nuc, being smaller than a regular colony and having perhaps suffered a certain amount of trauma and confusion in being set up, might be said to have a different set of expectations regarding a new queen. She should be accepted readily. After a week or so this nuc, frames and all, can be added to the larger hive. The queen will be accepted more readily now since she has been laying at a near normal rate and, further is surrounded and buffered by what have become her own family in the nuc.

Marking Queens

There is much to be said in favor of marking queens. The two obvious reasons for doing this are to help find her in the crowded and active conditions of the hive, and to keep track of her age. There is also much to be said in favor of individual beekeepers learning to mark their queens. Certainly if you buy your queens you can have them marked by the producer for a small additional fee. If you are the producer, however, learn how. It is not at all difficult.

The "paint." Several kinds of paint are suitable. Lacquer sold in hobby stores for model cars and planes, typing correction fluid from a stationery store, and fingernail polish will all work. Whichever is used, test it first to see that it dries quickly. Some of these materials lose their quick-drying ability as they age. Correction fluid and fingernail polish come with their own applicator brush, but model paints have a better range of bright colors.

You may use any color that you like but there is an international convention which specifies the color to be used each year in a five-year cycle. (Not all commercial queen producers follow this convention.) The purpose of this is to allow you to know the age of your queen by her color. The code is as follows:

> For year ending in: 0 or 5 blue
> 1 or 6 white
> 2 or 7 yellow
> 3 or 8 red
> 4 or 9 green

A queen marked in blue, for instance, was born in 1990.

Catching the Queen. In order to mark the queen she must be captured and restrained. This is not difficult though there is a certain amount of apprehension the first time or two that you do it. For marking there are two methods of capturing the queen. One involves picking her up; the other requires that you place a device over the queen on the comb to immobilize her. I prefer to pick her up. I am not comfortable holding her with a device that conceivably could injure her. In my experience she wiggles a lot and undue pressure is required to keep her still.

My method is to pick her up by the wings and thorax (never the abdomen) between the forefinger and thumb of one hand, and then to transfer her to the other hand so that she is held by two or three legs, again between the forefinger and thumb. Hold her by the legs of one side of the body only. If you hold opposing legs she is much more able to struggle.

MARKING THE QUEEN. The queen has just been marked with a dot of paint on her thorax. She is being held firmly between thumb and forefinger by two of her left legs.

Marking. You now have her held in one hand, perched on the end of your finger or thumb, with the back of the thorax exposed. Dab a small drop of paint on the center of the thorax, pressing with some firmness so that there is good contact of the paint with her body. Do this carefully so that the paint is only on the thorax. Hold her for a short time until the paint dries, then release her into the hive. Check beforehand to determine the drying time of the particular paint.

Be careful of propolis on your fingers as you handle the queen. A little propolis in the wrong place and she may get stuck.

Clipping Queens

Many beekeepers clip the wings of their queens. Some both clip and mark. Clipping has a different purpose than does marking. Marked queens are more easily identified and their age determined. Clipping can also be used to determine age, though only on a limited basis. Left wings may be clipped in odd-numbered years for instance, right wings in even years. Clipping is primarily done, however, as a swarm-prevention measure. A queen with her wings properly clipped cannot fly. Without a queen to accompany them, a swarm may emerge from a hive but will return when it discovers that the queen is not with them.

Clipping is done with a small pair of sharp scissors. The queen must be handled to do this, as with marking. Once the queen has been picked up and secured with her legs between finger and thumb, the wings on one side of the body can be clipped. Normal practice is to take about one half the length of both wings on the one side. There is no need to take more.

Tips on Queen Handling

Both marking and clipping can be done at the hive. It is simply a matter of finding the queen and proceeding as described. There is some chance, though, of having the queen fly away and perhaps get lost. Practice in picking up and handling bees is in order. Start with drones. Gain some proficiency in picking them up and marking them. Don't clip them, however. They need to fly.

An alternate method is to do the marking or clipping in the house, in a room with a bright, accessible window. Release the queen (or your practice bees) at the window. She will try to escape through the window and you can pick her off the glass at your leisure. If she flies off into the room she will return to the brightness of the window very quickly. This is a good method if you have received a new queen in a cage and you wish to clip or mark her before introducing her to the hive.

CHAPTER 5

MAKING INCREASE

M ANY BEEKEEPERS, PERHAPS EVEN MOST, ARE CONTENT with one or two hives. They resist any temptation or opportunity to add more. For others, the opportunity to add another hive is almost irresistible. Obviously, there is a limit. But before reaching that limit what are the methods by which you might set out to increase your number of colonies? Acquiring an overwintered colony from another beekeeper is one way. Capturing a swarm is also a possibility (see Chapter 3). Buying a package is another way, and is generally understood by everyone. Buying a nucleus hive (a nuc) is a possibility, as is making up a nuc or making a split from your own resources. Let's start with buying a nuc, and then consider doing it yourself.

Buying a Nuc

A nuc is a small colony of three, four, or sometimes five frames, all with drawn comb, containing honey, pollen, eggs, brood, adult bees, and a queen. Nucs are often available for sale from a dealer or producer or even from another hobbyist beekeeper. The nuc will prob-

NUC BOX. This four frame nuc is being kept on hand in the bee yard as a resource. It is large enough to maintain itself as a colony through the summer months, but small enough to be moved easily if needed elsewhere.

ably be contained in a nuc box which may be included in the purchase price or which may remain the property of the supplier. The contents of the nuc box are transferred directly into the purchaser's hive along with enough additional frames of comb or foundation to fill the box, and the colony is under way. As with any new colony, the nuc should be fed. If all goes well, the colony will develop rapidly and is likely to give a crop the first season if started in the late spring.

There are several important considerations for anyone who is interested in buying a nuc:

■ Prior to taking delivery, find out the background of the queen. You should expect a new, well-bred queen. She will be either confined in a queen cage or running loose on the comb. If she is confined there is little problem. You can control her release when you have the nuc settled in its new home. If she has been released, find out how long ago. There is a period of risk for a few days after a new queen has been introduced in a colony. Disturbances, as in moving or transferring the colony, could result in the queen being balled and killed by the workers. This is much less likely to happen in a nuc than in an established colony that is being requeened, but it is a possibility to be considered.

■ Have a clear understanding of what you are buying and if possible inspect the nuc before accepting it. The frames should all be drawn comb and substantially filled with brood, honey, and/or pollen. It is not unusual to be offered a nuc that includes a frame of foundation or perhaps a frame of empty drawn comb. It is reasonable for there to be some empty comb for the queen to lay in, but, a four-frame nuc that includes a full frame of empty comb, or foundation, is really a three-frame nuc. Know the going price for nucs in your area and be sure to inspect carefully if a cut rate is offered.

■ Expect the frames and comb to be in reasonable condition. You should not expect the equipment to be new since the frames are drawn from overwintered colonies and are at least a year old. But neither should you expect the supplier to use this as a method for culling old frames and comb. Don't go overboard on this, though. A conscientious supplier may give you a questionable frame because of its contents — that is, it may be well filled with brood or stores.

■ Was the parent colony (or colonies) from which the nuc frames were taken being medicated with Terramycin? There is nothing wrong with this but it is something you need to know so that if true you may continue the practice. Keep in mind that medication does not eliminate American foulbrood, it simply suppresses it. As long as the colony continues to receive medication every spring and fall there should be no problem, but if treatment stops foulbrood may appear.

Now that you know what a nuc is, consider making up one yourself.

Making Up a Nuc

First, the frames that make up the nuc do not all have to come from the same parent hive. You may mix and match. Assuming that you want to make up a four-frame nuc, look for one frame substantially filled with honey and perhaps some pollen, and three frames containing brood in various stages of development but heavy on capped brood. Here also include some pollen, especially if there is none in the honey frame. If the nuc is going to be kept in a nuc box for a few days there should be empty cells available in which the queen can lay. If it is going immediately into a regular hive body, then space for laying is not a problem. There are no ironclad rules here, however. It is not necessary that there be one frame of honey, for instance. What you should work towards is equivalent amounts. Three or four frames each with some honey will do if they add up to about as much as might be found in one well-filled frame. Similarly, pollen in each of several frames is fine, perhaps even better than in one. Generally speaking, pollen should be as close to the emerging brood as possible. It is the first thing that new bees look for after emergence. The equivalent of two and one half to three full frames of brood should be your goal. If you are making up a larger or smaller nuc, say three or five frames, then the quantity of each of the component parts should be proportionately more or less.

You have at least two choices as to the source of a queen when you make up your nuc. You can acquire a young, mated queen and install her when you make up the nuc. This is probably the best route. You will then have a laying queen almost immediately and your new colony should progress rapidly. Alternatively, you may give the nuc a frame of brood that contains eggs or day-old larvae, or you may give them a frame with an already developing queen cell, allowing them to raise their own queen. The disadvantage to letting the bees raise their own is the time lost in the overall development of the colony. The developing queen must first come to adulthood, which will take at least two weeks if the colony starts with an egg, and up to one week if they start with a capped queen cell. Several days will be lost while this new virgin queen matures and makes first her orientation flights and then her mating flights. Adverse weather can delay this even more. A few more days will pass before she lays her first egg and then a little more time until she is laying at full capacity. Total time from setting up the new nuc until the first egg is laid can be three to four weeks. If this is happening at the height of the season, as is most likely, we have a loss of potential new bees amounting to perhaps 1,000 to 1,500 per day, or many thousands for the full period.

To finish up your nuc, shake in some extra adult bees from the parent colony. There will be a certain number of adults, primarily house bees, on the brood frames that you transfer. There will probably be some older bees on the honey frames. If you are establishing this nuc in the same yard with the parent colony you can assume that some number of the older bees from the nuc will find their way back to the parent colony after their next foraging flight. To compensate for this, shake extra adults from the remaining brood frames of the parent colony. Be generous. The parent colony is well established and should make up its losses easily.

Making a Split

Sometimes you will come into spring with a real buster of a colony, one that you know is going to swarm if left alone. This hive is a good candidate for being split. There are at least two ways to make a split.

In its very simplest version, a split is made by taking one of the two hive bodies from a colony along with whatever its contents may be, and placing it on a new stand with its own bottom and covers. There are now two hives where formerly there was but one. One of these new colonies will be queenless and must either receive a new queen or raise one of its own. This simple method is followed often but success is far from guaranteed.

There is a more thoughtful method that does not require an appreciable amount of additional effort. The first step is to inspect the

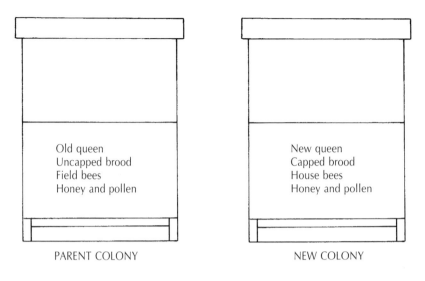

| PARENT COLONY | NEW COLONY |

Old queen
Uncapped brood
Field bees
Honey and pollen

New queen
Capped brood
House bees
Honey and pollen

FIGURE 5-1

colony carefully, identifying the frames by contents. A spare hive body is handy to use in sorting frames. Set up the new hive on its stand and in it place about half of the brood from the original colony. This brood should be primarily the capped brood. Shake in some extra bees from the uncapped brood frames. Add at least half of the honey and pollen. Introduce a new queen. You now have two colonies with contents as shown in Figure 5-1.

Queen introduction problems have been minimized by introducing the new queen to a colony containing almost entirely young bees. The older workers, the field bees, who are most likely to be antagonistic to a new queen, have been left on the old stand with their original queen. Any of the field bees who are transferred to the new hive during the splitting process will almost certainly return to the parent colony after their first foraging expedition. Extra bees were shaken into the new colony from the brood frames of the parent colony to compensate for this, and also to be sure that there are plenty of young bees to tend the uncapped brood.

The new colony will begin building population immediately as the capped brood emerges, and the new queen will start to lay almost immediately, as soon as she is released. In the parent colony the original queen, who has already proven herself by heading up a splitable colony, will continue laying as before.

This is a far more efficient way to split a strong hive and should result in two equally strong and capable colonies as the season progresses.

The foregoing has assumed that initially there is a single strong colony to be split. Perhaps though you wish to make increase but do not have a single colony that you believe is strong enough to be split. If you have two moderately strong colonies though you can use them to make up a third. Follow the same logic as in making up a regular split, leaving existing queens in their original colonies and those original colonies on their original stands. Take some capped brood from each of the two parent colonies, shake in extra adults from the brood frames that are to be left behind, give this new colony a share of the honey and pollen, and introduce a new queen. You now have three colonies where you formerly had two.

A question that is often asked is — won't the bees taken from two different colonies fight with each other when combined? They rarely do. There is too much confusion, and neither group of bees is in what it perceives to be its own home. Bees fight to defend their hive. By the time these bees have all come to accept this new home, they all will have acquired a common odor and accepted the new queen.

Minimizing Robbing

Whenever you manipulate hives and transfer brood chambers, frames, or supers from one hive to another you are bound to transfer some field bees. Assuming that all of the hives you are manipulating stay in the same bee yard or at least the same neighborhood after you are done, most of those field bees will eventually find their way back to their original location. Presumably with a new colony you are going to do some feeding. The field bees will be accepted readily in the new colony if they happen to return there, and the feed may be an enticement to them. They may start carrying it away, thereby initiating robbing.

Here are some steps that can be taken to reduce the possibility of robbing:

BUFFERED ENTRANCE FEEDER. A Boardman feeder is often an invitation to rob because of its easy accessibility at the entrance. Here, potential robbers must actually enter the hive and travel across behind the buffer to get at the feeder.

■ Bees are less likely to rob when there is a nectar flow underway. Make up splits and nucs only when there is a flow.

■ If possible, remove the new colony to a different yard immediately.

■ Delay feeding the new colony a day or two, until the initial confusion is past. This will reduce the attractiveness of the situation to possible robbers.

■ Reduce the entrance of the new colony so that it is more easily defended.

■ Do not feed with an entrance feeder. Since the feeder is in effect outside of the hive it is an invitation to robbers, especially if it leaks. If you must use an entrance feeder, use it in combination with a cut down entrance reducer so that the actual entrance is not directly beside the feeder.

Buying an Overwintered Colony

By keeping your ears open when you are among your fellow bee-keepers, there is always the possibility of hearing of a hive or hives for sale by someone who is getting out of beekeeping, or who suddenly finds himself with too many hives for whatever reason. This is a perfectly acceptable way to make increase but, just as with buying a nuc, there are cautions.

■ What is the condition of the equipment? Many are the times that a purchaser has gone to pick up a hive and found that it is only held together by inertia. The bottom board, or perhaps a brood chamber, has rotted out from neglect and the first attempt to move it causes it to crumble. Or perhaps the comb is old and black and contains an excessive amount of drone cells. These are not conditions that would necessarily cause you to reject a hive but they are certainly factors to be considered when agreeing to a price.

■ What is the age and background of the queen?

■ Has this colony been under a program of medication for disease or any other condition? As mentioned before, treatment for foulbrood must be continued.

■ Finally, do not bring home disease or mites. Insist on a recent inspection certificate, preferably dated within the past thirty days.

CHAPTER 6

LIFE OF A WORKER

BEEKEEPERS, INDIVIDUALLY AND TOGETHER, spend large amounts of time thinking about and talking about hive management and doing hive manipulations. Too often the emphasis seems to be on beekeeper activity and equipment, with not enough thought given to the bees themselves. The beekeeper does need to think about the bees — what they are doing and why they are doing it. A better understanding of the bees' lives will make colony management much more understandable and rewarding.

Each of the three types of bee in the colony — the queen, the drone, and the worker — has its own life and activities. These lives are totally intertwined and interdependent. For now, though, we shall only discuss workers.

A worker bee's life can be divided into three phases: brood, house bee, and field bee. The brood stage lasts for three weeks. Actually, it can be slightly longer or slightly shorter depending on outside factors such as the weather, or more precisely, the temperature. This variation is usually no more than a day or so either way, and for the most part it is not a factor in hive management. Beekeepers should be aware of it, though.

The other two phases, combined, are generally said to last for about six weeks — this is the figure most often seen in print and acknowledged by most beekeepers. In recent years we have come to realize that six weeks is excessive for the active season: four weeks is a better figure. During the off season, of course, it is much longer, and may for some bees be measured in months.

An individual bee begins life as an egg. After three days the egg hatches into a larva. The larva is fed continually by the nurse bees and increases in size enormously. After six days, feeding is discontinued, the cell is capped, and the larva becomes a pupa. During the next twelve days the pupa slowly transforms into a full-sized worker bee, emerging to begin her life as a house bee.

Many beekeepers mistakenly believe that adult bees grow larger as they mature. This is not so. All bees emerge from their cells, from their pupal cocoons, fully grown. However, for a few hours a newly emerged bee may appear to be smaller. Like many other insects, bees have an external skeleton made of chitin. When the bee first emerges the skeleton is soft, but within a few hours it hardens. Further, the hairs that cover the bee's body are somewhat matted at emergence. The hardening of the exoskeleton and the fluffing out of the hairs can make the bee seem larger.

Without question, there may be bees of different sizes in any colony, or from colony to colony. This is not a function of their age, however; it has to do with parentage, cell size, and hive conditions while they were developing. A colony that was recently started from a package, with a new well-bred queen, and that was hived on foundation rather than on old comb, may very well have larger bees than will an older colony. Presumably in the package colony the queen was bred with care from selected stock, and the breeder made some effort to control the mating situation so that the drones also came from selected stock. The cells in the hive, being drawn from foundation, and still relatively new, would be of maximum size. A package colony normally is fed by the beekeeper, so plenty of food would be on hand.

In an established colony the newly emerging bees could be just as large, but there is more chance for variation. Each time a brood cell is used, the emerging bee leaves behind its pupal cocoon. This cocoon adheres tightly to the cell walls and the workers are unable to remove it. Over time the cell becomes progressively smaller. Furthermore, an established colony is more likely to have workers of uncertain background as a result of swarming or supersedure. In such instances the colony will have a new queen who may or may not be of the same quality as her mother, and will have mated with whatever drones were available at the time. The quality of the resulting workers is always unpredictable.

Other conditions in the hive can also affect the size of developing bees. A colony that has been weakened by disease, drought, or a pesticide kill may raise undersized bees because of the resulting shortage of nurse bees or food.

The House Bee

Now that we have our new young worker out in the hive, what does she have to look forward to? Her life is largely pre-programmed, her activities being a function of her physical development and capability at any given time. An overriding factor though is the needs of the colony. This should become more clear as we continue.

We have considered the size of a bee when she first emerges from her cell as an adult but we have not yet considered her capabilities. A newly emerged bee is limited. Though she has reached her full size, her muscles and glandular systems are not fully developed. In a normal colony these will continue to develop during her life as a house bee, and this development will control the activities that she is able to carry out.

There is often mention among beekeepers of the division of labor within a hive. This is not as strict a division as some believe, as is found with some other types of insects. Every worker bee, at some time in the course of her life, has the capability to perform every job in the hive, though in reality she will not. The division of labor in a bee hive is more a matter of dividing each bee's life into segments during which she is capable of doing certain tasks. After that segment of her life has passed she is no longer capable of performing that particular task. As stated, the capability is related to the muscular and glandular development of the bee and this development takes place in a set and orderly manner in the normal course of events.

The possible house bee duties or activities for an individual worker are many. They include:

- cleaning cells
- tending the queen
- patrolling the hive
- building comb
- processing nectar

- tending brood
- resting
- capping cells
- packing pollen
- guarding

When a worker emerges from her cell, the first thing she does is eat. She will seek out pollen, her source of protein, to give her body the material it needs to complete muscular and glandular development. She will solicit honey or nectar from older workers, to give her the carbohydrates she needs for day-to-day living. Very quickly thereafter, within her first hours of adult life, she will begin work. Her first days are spent in the darkness and warmth of the brood nest and her first task will most likely be the cleaning of cells from which she and other young workers have recently emerged. It is a relatively simple task that requires a minimum of experience and capability.

This first task, or any task that a house bee undertakes is not exclusive. She may alternate between several tasks in a given period. Nor does she work all the time: she spends as much as a third of her time simply resting or sleeping. Another third of her time may be spent in patrolling the hive, looking for work and assessing hive conditions. If in her meandering she discovers work needing to be done, and if that work is within her capabilities of the moment, she will probably undertake it. If, for instance, this young bee discovers a newly vacated cell that needs to be cleaned, she may do it. When that is done she may find another cell to clean or she may continue patrolling. With only one cell being vacated by each newly emerged bee, no one bee will necessarily do a lot of cleaning, though each cell may be cleaned several times before it is actually reused. It may take a while before the queen comes back to a particular cell and it must be kept prepared.

In the course of her wandering, which at this stage of her life is confined to the general area of the brood nest, our bee may encounter the queen. She may briefly become a part of the queen's court, cleaning and grooming the queen as she moves about the hive. Here our bee is doing her part to spread the message that the queen is present and that all is well. She does this through her physical contact with the queen. As she grooms and cleans, the worker picks up queen substance from the queen's body. Her participation in the queen's court is a passing thing, measured in seconds or minutes though she may have many such encounters during her time as a house bee. After she breaks off from her contact with the queen, the worker will continue to circulate through the hive, patrolling and working and inadvertently passing on traces of the queen substance to each bee she encounters. The continued cycling of bees into and out of the queen's court keep the queen substance and its attendant message of well-being circulating continually through the hive.

During this initial period of her life, though our worker has stayed largely in the brood nest, she has ventured out into the world briefly. Sometime towards the end of her first week she will probably make her first orientation flight, joining in the so-called play time at the hive entrance. Periodically after that she will make further such flights, preparing for the day that she makes her first foraging trip. Each of these flights will be relatively brief and will not take her far from the hive.

Back in the hive, her varied life continues. During that first week her brood food glands begin to function and she is able to feed the developing larvae. She also begins to manipulate wax, aiding in cell building and capping. Later her own wax glands will begin to function and she will secrete wax and make a major contribution to comb building. Hive cleaning is an ever-present need and periodically she will participate, carrying out discarded cappings, dead bees, and other debris. Meanwhile, she is slowly working her way out of the brood nest.

During her second week our bee may be found on the periphery of the brood nest where she will pack the pollen that has been deposited in cells by the incoming foragers. Moving a little farther out, she may be found in the honey storage area where she will receive nectar from incoming bees, processing and storing that nectar as a step in its conversion to honey. She is slowly working her way to permanent duties at and beyond the entrance. She is on the verge of becoming a field bee.

During the transition from house bee to field bee, there are two tasks she may perform at the entrance, fanning and guarding. Fanning of course is an aid in ventilating the hive. It might be considered rather mundane work but it does depend on well-developed wing muscles, a necessary prelude to serious foraging. Guarding, the final task before graduation, takes place at that point in her life when the sting glands are well developed, giving maximum amounts of alarm pheromones and venom.

The Field Bee

Finally the great day has come, and life as a forager begins. The general belief has been that the foraging period of a bee's life lasts about three weeks. Increasing evidence has shown that a bee's total adult life span is closer to four weeks during the active season, and that life as a field bee is sometimes as short as four or five days. A bee's life span is tied to the metabolization of carbohydrates. The more work that she does, the more carbohydrates (sugar) she consumes and the faster her system wears out. Foraging is the most energy-consuming activity a bee can undertake. Of course, some bees do live for six weeks and longer, and forage over a period of three weeks and longer, even in the summer. Bad weather, poor nectar flows, and anything else that forces the bees to be idle contribute to a longer life. Unfortunately, these same causes result in a poor season for the beekeeper.

CHAPTER 7

SUPERING

BEFORE DISCUSSING SUPERING we should define a super. Many beekeepers, and some beekeeping books, talk as if every box on a hive is a super. Others consider that supers are boxes of a certain size, usually shallow or medium depth. Discussion between beekeepers can become very confusing when there is no common definition. Here we will accept the definition that a super is part of the superstructure of the hive, the part above the main hive or hive bodies. We will assume that the standard basic hive comprises two full-depth (9⅝") hive bodies, recognizing however that this is by no means a universal configuration. The two hive bodies may also be termed the brood chamber and the food chamber. Above these hive bodies are the supers, and they may be any depth. The four standard depths are: comb (4¾"); shallow (5¾"); medium (6⅝"); and deep (9⅝"). (See page 61 for more on sizes.) It is the position on the hive that determines if a box is a hive body or a super. For our purposes, any box in the third story or above is a super, no matter what its size.

When discussing supering, some regularly heard questions from novices and more experienced beekeepers alike are: how should we super, top or bottom? When should we super? How many supers should we put on? All at once or one at a time? And more. The basic answer to all this is — put on as many supers as the colony needs at the time that they need them, and whether they go on the top, the bottom, or in the middle is often not important. This, of course, is not the kind of answer that most people want to hear. They want a specific and precise answer. However, beekeeping should not be done by rote. It should be done with an understanding of what is being done and of the implications of each action. With this in mind, here are some considerations:

■ The contents of the super are important. That is, does your super contain foundation or drawn comb? Bees do not draw out foundation unless there is a moderate to strong nectar flow in progress (or they are being fed). They are not capable of secreting wax unless nectar or syrup is coming into the hive. The continuing engorgement of the bees as they receive and process the nectar in the hive stimulates their wax glands to activity.

■ Bees are sometimes slow to draw out foundation above a queen excluder and this is especially true if there is no drawn comb at all above the excluder. This is even more true when the nectar flow is weak. The house bees are not sufficiently stimulated by the weak flow to secrete wax. Baiting that super of foundation with a drawn frame or two, especially if those frames contain some honey, may help in a marginal situation to bring the bees up.

■ A colony should never be allowed to develop a sense that it is becoming crowded or is running out of storage space. Crowding is one of the factors that can lead to swarming.

■ The average moisture content of nectar is 30 percent to 40 percent. Some nectars have as much as 70 percent. Honey has less than 19 percent moisture. The reduction in moisture takes place in the hive over a period that may be several days. There must be room in the hive to store the extra bulk of the nectar while it is being reduced and ripened.

CHEWED FOUNDATION. **This foundation was placed in a hive too late in the season. Though partly drawn in some places, it has had holes chewed in it and is propolized.**

■ It is better to put on too many supers too early than to put on too few too late. Nectar that was available but was not gathered is gone forever.

■ When there is a good nectar flow underway, the bees will find the empty supers whether they are placed on the top, the bottom, or in the middle of the stack.

■ Empty supers left on the hive during a dearth of nectar will very likely be propolized and travel stained, and the wax may be chewed around the edges. Damage to the comb or foundation from chewing can often be extensive as the bees take the wax to use elsewhere in the hive. With extracting frames we can live with this though it certainly is not something to encourage. The bees will ultimately repair any damage and use the comb. For comb honey frames, however, this chewing is not desirable. Good comb honey is made with fresh, tender wax that is free from propolis. Keep those comb honey supers off the hive until you know that a nectar flow is in progress and the bees are working.

■ Schemes for supering that involve manipulating and changing the positions of all the supers every time one is added are probably not worth the effort. Beekeepers who have success with such schemes are probably good beekeepers who would have success with any system.

Timing

Definitions and such aside, when do you put on supers? There are schemes and there are schemes but, at least to begin with, a very simple approach is best. The simplest is to super when the first significant nectar flow of the season is underway, when it is apparent that honey is being stored in the upper hive body. This requires that you watch the hive closely during this period. Any delay in supering could encourage swarming.

Once the additional super is on, place additional supers as needed. This latter is a rather vague statement, but the timing is quite variable. With a strong nectar flow I like to get the second (or third) super on when the first (or second) is one-third to one-half full. If it is a weak flow I may wait until it is perhaps three-fourths full. I prefer to end up with a few supers completely full rather than with several supers only partially full. Supering is a balancing act. Proficiency should come through experience. It is usually best, though, to have too many supers on a hive rather than too few, especially at the beginning of a nectar flow.

Placement

The next consideration is where in the hive stack to put each super, assuming that you add one super at a time over a period of time. Some beekeepers put all of their supers on at the beginning of the season and avoid any of these questions. This system is acceptable, though it can have some drawbacks. One is that the beekeeper may be less inclined to pay attention to the colony if all the supers are in place. There is less incentive for inspection. In a weaker hive, a large volume of empty space is a possible place for wax moths to get in and cause problems.

We will assume here that the supers are going on one at a time. Placing the first one is no problem. There is only one place to put it. With the second one you have a choice: it can go over (top supering) or under (bottom supering) the first. If the super to be added contains drawn comb it can go on top. If it contains foundation it is probably best to put it under, "baiting" it with one or two combs containing honey taken from the first super. This encourages the bees to start working that super. If a super of foundation goes on top, or is not baited, the bees may ignore it if the nectar flow is not strong or is approaching the end. This could result in excess crowding below as the bees build burr comb and stuff honey in every possible space in the hive bodies and lower supers. When placing a third super, use the same procedure, putting supers with foundation under one or more filled supers or baiting them. Always keep in mind that it takes a strong nectar flow to get foundation drawn. The bees need the stimulation of a nectar flow to get the wax glands going.

Supering Strategies
A basic scheme or method for supering is simply to put on one or two supers at the beginning of the season, keep an eye on them, and, when the second super is one-half to two-thirds full, add another. As each successive super becomes two-thirds full, add another. In each instance, top super. Full supers may be taken off and extracted or left on until the end of the season. This may be all of the system or scheme that you will ever want; for many it is sufficient. However, many books and other publications on beekeeping have advice on supering. Read other beekeeper's ideas and methods. Try different schemes. Ask questions of other beekeepers. Decide what works best for you. Recognize that everyone has different goals, attitudes, amounts of time available, and so on. Nectar flows vary from region to region and to some degree from year to year. Not all methods work the same for everyone nor will everything work exactly the same every year.

What Size Super
Four basic sizes of honey supers were mentioned earlier. There are other sizes, though they are no longer at all common, and they probably should be given no consideration by the average beekeeper.

In your first few years of beekeeping it is always best to stick with the standards. There is no harm in experimenting, in being innovative, but only after building a good base of experience. Competence in beekeeping is not gained in one or two years. With this thought in mind let's look at the standard sizes.

■ *Deep — 9⅝"*. One advantage to using deep (also called full-depth) honey supers is that they are the same size as the hive bodies. You are then dealing with only one size of box for the entire hive. Without question, this is a significant advantage, giving interchangeability and flexibility. It is also a more efficient size when extracting. Essentially the same work is done in extracting a large frame as is done in extracting a smaller frame, but fewer frames need be handled per volume of honey. The disadvantage is the weight. A deep super when full will contain upwards of 50 pounds of honey. The total weight of the full super, both honey and equipment, can approach 70 pounds in a good season. Considering the shape and size of a super and the conditions under which you frequently work when taking off honey, this weight can be unacceptable to many people. The deep super is a size favored by many commercial beekeepers who are concerned with efficiency, and who often do not have to do the heavy lifting themselves.

■ *Medium or mid-depth — 6⅝"*. The medium depth is a compromise between the excess weight of the full depth super and the more limited capacity of the shallow. It holds about 35 pounds of honey, a much easier load to handle. It still can be awkward for some people.

■ *Shallow — 5¾"*. This is probably the most common size among hobbyist beekeepers, primarily because its lighter weight makes it easier to handle. The shallow holds about 30 pounds of honey. It can also double as a comb honey super for producing full frames or cut comb.

■ *Comb — 4¾"*. This size is intended for comb honey production, using round or square section equipment. There are also frames available to be used in producing full frames of comb honey, but these frames are becoming increasingly difficult to find. This is not a size ordinarily used for extracted honey since the frames that are available are specialized for comb production.

If forced to make a recommendation I usually suggest that hobbyist beekeepers choose the shallow size. The difference in honey capacity is not great between the shallow and the mid, and the smaller size and less total weight make the shallow noticeably easier to handle. This is even more important as the years catch up with us.

I suggest that you do not have both shallow and medium depth supers in your operation. Choose one or the other. With both on hand, the frames somehow keep getting interchanged. Shallow frames in mid supers leave empty space at the bottom of the supers. This leads to burr or wild comb on the bottom of those frames if not discovered in time. Mid frames in shallow supers are usually obvious since they are about an inch too tall, but it is possible for one to sneak by, not becoming obvious until you try to put the super on the hive. This may not be a major problem, but why bring on any problems at all if you can help it.

How Many Frames?

When supers contain only nine frames (some say eight is even better but this is questionable), the cells are drawn out more deeply, and the frames are without question easier to uncap than if ten frames had been used. Further, ten frames drawn normally or nine frames drawn deeply hold about the same amount of honey. However, don't start with nine frames of foundation. Though they may be equally spaced in the super there is more space between them than normal, and too often the bees will fill some of that extra space with wild comb before or while drawing out the foundation. When using foundation, start with ten frames. After they are fully drawn, reduce to nine. The bees will draw each of the nine a little deeper. It is usually most convenient to make this reduction the following year when the supers are being placed for the new season. Store the tenth frame away as a possible replacement for the future.

Frame Spacers. Nine-frame spacers are available for permanent installation in honey supers. These are notched metal frame rests that are permanently fastened in place. These spacers guarantee that when

ABOVE: Nine-frame Spacer in Super. This nine-frame spacer is permanently attached in the super. Lateral movement of the frames is restricted, and the super now cannot be used with ten frames.

BELOW: Nine-frame Spacing Tool. The nine-frame tool is easy to use and allows for more flexibility in the use of the supers.

only nine frames are used they are equally spaced and do not slide out of position. Many beekeepers use them. I do not care for them. Once in place they are a permanent part of the supers. You are then forced always to use nine frames. This is sometimes not desirable, as for instance when you are putting in frames of foundation. Spacers also limit the movement of the frames if you want to slide them sideways to help in removing them. I prefer to carry a nine-frame spacing tool with me. The minor inconvenience of the extra piece of equipment is more than outweighed by the flexibility it gives in using the supers. Further, I have found that the bees will very quickly build their own spacers by filling between frames with propolis so that most of the time the tool is not necessary.

CHAPTER 8

HIVE INSPECTIONS

∎

A COMMON QUERY FROM MANY BEEKEEPERS relative to hive inspection is — what am I looking for? A surprising number do not really know, and this applies to many experienced beekeepers as well as to novices. They blindly open and pull apart the hive, frame by frame, with no purpose in mind. Others have been told that this is the proper way to manage a hive. They have been told that they should open the hive every week or perhaps every two weeks, pull out every frame and inspect it closely, find the queen, cut queen cells, check the brood for disease, analyze the brood pattern, and scrape burr comb and propolis. Such intense management, or perhaps misman-agement, will not lead to a thriving colony or to a bumper crop.

This leads us to a good rule — always have a purpose or goal in mind when opening a hive or pulling a frame. However, it is not always necessary to open a hive to inspect it. A great deal can be learned from external inspections. The habit of making such inspections should be cultivated. On a given visit to your hive an external inspection may be all that is necessary.

External Inspection

A large amount of information is available simply from watching the activities at or near the entrance of the hive. This is often the basis of the so-called "let-alone"[*] school of beekeeping. A competent beekeeper, whether of that school or not, should cultivate the ability to watch and analyze, using the observations both to supplement and, sometimes, to substitute for regular internal inspections.

Of course, this works only if you are starting with a knowledge of the appearance and activities of a normal, healthy hive under a variety of conditions, and only if you have the experience and the confidence to interpret properly the information that is available. Developing the eye to interpret external signs is simply a matter of observing, relating everything you see to internal conditions once you open the hive, and, ideally, keeping notes. External inspections can be even more effective if there are two or more hives in the bee yard to give a basis for comparison. Certain problems may not be obvious in a single hive, but if there are two hives, a problem often stands out by comparison. If your hives are close to your home and you are able to watch them frequently, even daily, your diagnostic capability will develop rapidly.

Assume that you do have at least one normal, healthy hive, it has come through the winter nicely, and you have given it a thorough spring inspection and cleaning. What can you determine from external inspection as the season progresses? There are a number of specific activities to look for at the hive entrance. The absence or presence of any of these activities, alone or when coupled with each other, with the weather, with the time of year, and with the activities of other nearby hives, can tell volumes to the enlightened beekeeper.

First, though, a weight check is in order. On first approaching your hive, go to the rear, reach down, and lift the back end an inch or two. Lift it only high enough to get a sense of the relative weight of

---

[*] *"Let-alone" does not means ignore. It means to minimize contact and inspections of the hive on the assumption that the bees, having survived for hundreds of thousands of years without our intervention, can continue to do so with only a necessary minimum of intervention today. Successful let-alone beekeeping is based on a thorough knowledge and understanding of honey bees. It is not a recommended practice for beginners or novices.*

the hive. Be sure that you are lifting only the hive and not the hive stand. The first few times you do this may tell you little, but over time, and when coupled with your further observations and inspection, this relative weight is an important bit of information. For instance, a hive that feels light in the middle of a major nectar flow is an immediate cause for concern. A hive that feels light when others around it are noticeably heavier is cause for concern. It is a signal to check further in that particular hive.

Once you have performed your weight check, then go on to make more specific observations.

Traffic Volume

Obviously, traffic volume is not going to be the same from day to day or from week to week. There are times, though, when you should or should not expect certain volumes. On a beautiful spring day, for instance, with fields of dandelions in full bloom, it is reasonable to expect heavy traffic, with large amounts of nectar and pollen coming in. On such a day, limited traffic in and out of a supposedly healthy, normal colony is cause for further inspection.

On the other hand, once you have a handle on the year-to-year pattern of nectar flows in your area, you should not be surprised to see a very low volume of traffic at certain times. In parts of New England, for instance, there is usually a dearth of nectar from about mid-July until mid-August. It is normal for there to be very little foraging during that time. Further, some crops do not secrete nectar all day. Buckwheat, for instance, yields nectar in the morning but not in the afternoon. This means that to supplement understanding of the bees, the beekeeper must have a knowledge of what plants are yielding in a given period and the characteristics of those particular plants.

Other conditions that can affect traffic volume are temperature, swarming, and pesticide losses. Temperatures in the nineties inhibit foraging. The bees stay home, often clustering heavily at the entrance. Many beekeepers see such clustering as a prelude to swarming. It is not: it is simply too hot to work, and too hot and crowded in the hive. In a swarm situation there will be that brief period of a day or two when foraging is reduced as the colony prepares to swarm, but the masses of bees are not likely to be hanging on and under the entrance.

They will be restless and moving about. After the swarm has departed and the home folks have settled down, there will be a smaller foraging force, though this can be difficult to detect.

With pesticide losses of any severity, foraging will be reduced. Sometimes the cause will be obvious, as dead bees pile up at the entrance. Occasionally, though, the cause will not be obvious, if the pesticide is one that kills foragers out in the field. They do not return to the hive to die and so their loss is not immediately evident.

A caution is in order here. Be sure that you recognize the difference between a heavy, normal, traffic volume and the heavy traffic resulting from massive robbing. The former is orderly (within limits) and deliberate. You will come to recognize normal traffic, whether heavy or light, by watching regularly. Robbing behavior is quite different, especially in the early stages when the hive is defending itself vigorously.

Robbers

There are two types of robbing activity. One type might be considered active, the other passive. In the first, a bee or perhaps several bees discover a soft touch, a weakened hive that is lightly guarded and they endeavor to break in. Such robbers have a distinctive action. They tend to be sneaky. They do not approach a hive entrance directly but ease in from the sides. Their flight pattern is somewhat erratic as they bob about looking for a way in. If challenged they usually back off but keep trying. Eventually some of them may get in, find some honey to steal, return home and recruit more of their hive mates. As the numbers of robbers build up, they are more deliberate in their approach, and fighting becomes evident as they force their way in. The robbers may be successful, or they may be repelled and the incident is over.

If the robbers are successful, they overwhelm the defenders and proceed to take all of the honey back to their own hive. Here we are in a passive type of activity. No further defense takes place. The defenders are completely demoralized and do little or nothing. The robbers continue until nothing is left. It is this type of activity that may be mistaken for heavy foraging with continual traffic in and out of the hive. If there is any question, open the hive and inspect the honey storage areas. Bees may be found actively removing honey from cells. The opened cells will have ragged rims where the robbers have torn them open instead of neatly trimming away the cappings. The bottom board may be littered with the cappings.

It is this type of activity on a nice day in late winter or early spring that may cause the beekeeper to believe that a hive is alive and well, when actually it has died out and is being robbed of its remaining honey.

Aggressiveness

Well-bred bees are not normally aggressive. Certainly they will act aggressively at times, if, for instance, they perceive a threat to themselves or their hive, but treated with due respect they are relatively passive. A colony that has suddenly become aggressive may catch you by surprise some day. It is easily corrected once the cause is determined. Assuming that your colony was normally non-aggressive, why is it suddenly a menace to you or the neighborhood? Possibilities include regular visitations by skunks or vandals, an unsuitable location, or a queen of questionable breeding. A little analysis should reveal the cause. Are there signs of skunk predation in the beeyard? (see photo next page) Are there signs of human mischief? Have overhanging trees or shrubbery just leafed out, putting the hive in heavy shade? Did the colony swarm earlier this season, or supersede their queen? Either condition could result in a new queen of questionable lineage whose less desirable offspring are just now showing up. Any of the foregoing situations could make the colony more aggressive. Any of them should be easily detected with regular attention to the hive.

Guarding Behavior

Guarding is a normal part of hive activities, although the number of guards and the intensity of their actions varies through the season. Generally speaking, the more intense the nectar flow, the less intense is the guarding. When a good nectar flow is on, all the colonies in the area are preoccupied with gathering the available nectar. It is when the nectar flow is down and foragers have time on their hands that robbing may become a problem and the guard force is strengthened.

Skunk Activity. A skunk has been active here. Note the bare earth, the hole, and the matted grass. Sometimes the entrance board will be streaked with dirt or mud.

If your colony is guarding when there is a strong nectar flow, or not guarding during a dearth, investigate hive conditions carefully.

Loafing

Numbers of bees hanging out at the hive entrance can indicate several things: a dearth of nectar, weather too hot to forage, or no place to store honey. You cannot do much about the weather, but you can provide either temporary or permanent shade as appropriate and perhaps raise the outer cover a little to provide ventilation. If the problem is a dearth of nectar, ensure that they do have food reserves on hand. If not, feed, but be careful when feeding not to set off robbing. If it is simply a matter of inadequate storage space, then add supers.

Orientation Flights

At some point young bees must learn their way out into the world and back again. They start by taking short orientation flights. Often on a nice day, about mid-day, bees may be seen flying in large numbers about the entrance to the hive. If you look closely you will see that each bee has flown out, turned around, and is hovering and circling as it faces the hive. These bees are studying the appearance of the hive and its immediate surroundings so as to be able to identify their home when returning from a flight. This is the activity that so often strikes fear into the hearts of novice beekeepers the first time that they encounter it. Because of the mass of bees in the air, the immediate assumption is that the colony is about to swarm. This activity, often called playtime, is a sign of normal colony development.

As beekeepers we understand that bees are welcome only in their own hive, and that interlopers are quickly identified and thrown out. At least, that is what everyone says. But it is not entirely true. In beeyards with several or many hives of similar appearance that are placed in uniform rows, the bees may become confused. It is not unusual for bees to return from foraging and enter the wrong hive. If they are confident in their actions, landing and walking in deliberately, and if there are guards present, the trespasser may be challenged but allowed to pass. In fact, if the bee believes herself to be at home she may ignore the guard entirely, barging right on in. On future foraging trips this bee may return to her original hive, or may continue to orient to the new hive. When hives are placed in uniform patterns, it is important to place a few landmarks or mark the hives so that the bees can more readily identify their own. In extreme cases, large numbers of bees may drift to other hives, leaving some hives underpopulated and others overpopulated. This is particularly true with a long straight row of hives. Bees from hives in the middle of the row tend to drift to the outer hives, creating a significant population imbalance.

NOTE: Marking hives with color or designs is not a straightforward operation. Bees recognize only a limited number of colors and a very limited number of shapes or designs. Anyone planning to use hive marking to aid the bees in orientation would benefit from reviewing some of the literature. (See Appendix B, *Bees: Their Vision, Chemical Sense, and Language*, by Karl von Frisch.)

Internal Inspection

Though a great deal can be determined by observing activities at the hive entrance, this should not be the only type of hive inspection. It should be supplemented with periodic inspections of the hive interior. This does not mean tearing down the hive completely and examining every frame. That would be unnecessarily disruptive. Sometimes it is as simple as checking the status of the honey supers. Is it time to add another? Sometimes it means inspecting one or two selected frames from each hive body, continuing with more frames if anything seems suspicious. Look for an amount of brood that is appropriate for the time of year. Look for a proper balance of the different stages of brood — eggs, larvae, pupae — for the time of year. Look for a good, solid, disease-free brood pattern. It is not necessary to look specifically for the queen if there is evidence of her presence (eggs, young larvae), but know the signs of a queenless hive. Know also the signs of laying workers. Know the visual characteristics of the various diseases. And finally, look for an adequate amount of honey stores for the time of year. Before we can check any of this, though, we must open the hive.

Opening the Hive

The internal inspection should begin as you open the hive. Use as little smoke as possible so as not to mask the normal activities. Listen as you lift the covers. Does the voice of the colony rise, and then subside, indicating a normal situation, or does it rise and remain high indicating annoyance or a problem? Listen, and learn the voice of the colony in normal and abnormal situations. Learn also to distinguish the voice of an individual bee as it flies about your head. Is it annoyed or is it just checking you out? Learn the voice of a drone as compared to that of a worker.

After the covers are off, look. Do the bees return quickly to their chores after the initial disturbance, or do they keep running? Do they line up, heads up, between the top bars of the frame, apparently watching you? If so, be careful. Perhaps a little more smoke is in order.

As you open the hive are there any questionable odors? Normal hive odors tend to be pleasant: they are compounded from honey, wax, evaporating nectars, and the bees themselves. Long confinement can bring about moldy, musty odors. Those from diseases, such as foulbrood and dysentery, are sour and unpleasant. Experience will help you to distinguish each of these. Now it is time to get down into the hive and note specific activities.

Each time you open a hive you should have a specific goal in mind. It may be as broad as a complete spring inspection and cleaning or as narrow as checking the condition of the supers. In the former you will no doubt go right down to the bottom board; in the latter you will probably not disturb the brood nest at all unless you discover something as you work that leads you on.

If you intend to inspect both brood chambers then it is usually best to work from the bottom up. Set the supers aside on the cover, and then the second hive body. You now have both hive bodies exposed allowing for comparison and manipulation. Complete your inspection of the bottom hive body. Inspect the second hive body and the supers while they are off the hive. This way you are not further disturbing the more aggressive bees, the field bees and guards, who are oriented to the hive body on the bottom board. You also are returning the hive body and supers to the hive stand after you inspect each one, which helps to pacify them after the disturbance. The alternative, inspecting from the top down, causes you to disturb each box in order and then set it aside off the hive stand, where the bees can become further agitated as you complete your inspection.

Pulling Frames

It is difficult to remove frames from a busy hive body without rolling and crushing a few bees. You can minimize this damage, though, and it is in your best interest to do so. It does not take much in the way of angry bees to set off the entire hive and your inspection time can become very unpleasant.

After you have removed the covers, allow the bees to settle a bit. Choose the frame you wish to pull first. Ideally it will be next to the wall of the hive and will have the fewest bees on it. Be sure that it is not overly fastened to the hive wall with bridge comb. If it is, then it may be best to move to the next frame. Use your hive tool to gently pry the frame loose from the hive wall on one side and the neighboring frame on the other. Once it is freed, lift the frame slowly and carefully. These outer frames will usually contain honey. Look it over and then set it down carefully outside the hive. Stand it on end, leaning against something solid like the hive or the supers you have stacked on the bottom board. Now you have room in the hive to slide the next frame over so that it can be removed with minimal damage to bees and comb. Leave the first frame, and perhaps even the second, out of the hive until you have completed your inspection of the whole box. This gives you plenty of room to work, with the least disturbance and harm to the bees.

Pay attention to the orientation of the frames so that you can replace them in the same position and facing the same way as you complete your inspection. The bees have everything arranged as they want it. You should rearrange the orientation of frames only for good and specific reasons. For instance, you might choose to relieve congestion in the brood nest by moving an empty frame or two into the brood area.

CAUTION: Young brood is very sensitive to the drying effects of direct sunlight. Even brief exposure can be harmful. If you remove brood frames from the hive, and set them aside to continue your inspection, place them in the shade.

Queenless Hive

One of the most unfortunate events in the life of a colony is for it to become queenless with no resources available from which to raise a new queen. This problem may be compounded by the inattention of a beekeeper who is not aware of the situation and allows laying workers to take over. Queenlessness can come about in several ways: beekeeper carelessness in working the hive; cutting queen cells at the wrong time in a swarming situation; the queen lost on a mating flight after supersedure or swarming; or disease.

It is important to be able to distinguish between a queenless hive and one that is in transition. By queenless we mean a hive that is hopelessly so, with no adult queen, no active queen cells, and no eggs or young larvae with the potential to become queens. A hive in transition is one wherein there are queen cells started, or there is a newly emerged adult queen who has not yet started laying. During this period the bees recognize the situation and are perfectly content. They will resist any efforts by the beekeeper to introduce a new queen. You may test the status of a hive that you suspect to be completely queenless if you have another colony or a nuc on hand. Take a frame containing eggs or very young larvae from the second hive or nuc and place it in the colony suspected to be queenless. If they begin to raise one or more queens you can assume that the colony is queenless.

Laying Workers

If a hive becomes hopelessly queenless, we can expect that some of the workers will develop the ability to lay eggs. Usually this is a holding action with no positive outcome. Because the laying workers have no ability to mate, their eggs will not be fertilized. These eggs will result only in drones. Eventually the drone population will overwhelm the colony as the worker force dies out through attrition. The non-productive drone population will eat the available stores, and the colony will starve.

If you look carefully at the brood cells, the presence of laying workers is usually obvious. The brood pattern becomes very spotty, more than one egg or larva may be found in a cell, and the eggs are not always at the bottom of the cell. Those cells which are capped are domed or bullet shaped, elongated to accommodate the extra length of the drone.

When the presence of laying workers has become obvious, and especially when the evidence is in the form of mature, undersized drones, it is apparent that the hive has been queenless and perhaps neglected for some time. Laying workers do not develop in a hive as long as a queen or worker brood is present. Pheromones given off by the queen or the brood inhibit the development of the ovaries in workers. If brood from a laying worker is present it usually means that the colony has been queenless for four to five weeks, and often longer. If any of that brood has emerged as mature (though under-sized) drones, it means that the colony has been queenless for much longer.

If you discover that you have laying workers in your colony you must take positive steps to correct the situation. If you determine that the laying workers have been active for a relatively short time, as evidenced by only a few capped cells, you may be able to save the colony by requeening. If they have been active for a longer time and you are seeing many capped brood cells, along with mature but un-dersized drones, it may be too late for a rescue effort to be worthwhile.

Requeening a colony with laying workers can be difficult since the workers consider themselves to be queenright, and it is virtually impossible for the beekeeper to identify the laying workers to remove them. Requeening via a nuc is the method most likely to work. The nuc should have a well-established queen and a good work force. The newspaper method of combining may be used but I have been suc-cessful by simply inserting the frames of the nuc in the hive to be requeened, off to the side a bit if there is an area that is less busy. This gives the nuc occupants an opportunity to become oriented and es-tablished with little or no conflict with the original bees of the colony. With worker brood and a queen now present, no new laying workers will develop and the existing ones will be suppressed by the colony.

If it is apparent that the laying workers have been present for a long time and there are few workers left, it is probably not worth-while to try restoring the colony, especially late in the season. There will not be time for them to build up and prepare for winter. If you have another colony that is of reasonable strength, you may consider combining the two. If you have several colonies, then the best action may be to split up the salvageable parts of the laying worker colony among the several others. A final solution might be to let the colony die out naturally, then store the equipment away for future use.

Honey Stores
Beekeepers often seem surprised to discover that a super that was full of honey is suddenly empty. During the early season they see bees working, supers filling, and everything seemingly going well. Later in the season, as they begin to think of all those pounds of honey that

they will be taking off soon, they suddenly discover that some of those supers are not so full any more. This is not an unusual situation. Nectar flows are not predictable. A previously dependable crop may not yield that year. Drought or excess rain may wipe out particular crops or inhibit normal foraging. Meanwhile, life goes on in the hive. The bees must eat. Brood must be fed. Honey stores are consumed. Supers become empty.

An eye to the ongoing weather and the condition of bloom of the nectar plants of the area should help you remain aware of what may be happening.

Watch the Comb
In those new hives started this year on foundation, be sure that all of the frames are fully drawn well before the end of the season. Watch the outside frames. If they are not properly drawn, move them towards the center of the hive. Supering too soon with a new colony that was originally hived on foundation can result in the bees moving up into the supers without fully drawing the outer frames in the hive bodies. When fall extracting time comes and the supers are taken off, those combs in the hive bodies may still not be fully drawn. It is then too late to expect it to be done for that season. Successful wintering can be certain only if all of those frames have been drawn and are full of honey.

In Summary
Inspect regularly. In general terms, know why you are inspecting and what you are inspecting for each time you go to your hive. Think about what you see each time you inspect. Relate your findings to what you saw last time you inspected, and to what you saw a year ago this time. A record or journal will help immensely.

Remain in charge. Requeen periodically to maintain a young, vigorous queen in the colony.

CHAPTER 9

COMB HONEY

███

ONE OF THE COMMON MISCONCEPTIONS OF BEEKEEPING is that producing quality comb honey is a simple operation, one that anyone may do, even a beginner. It is often recommended to beginners as a way to get around the lack of an extractor. A second and related misconception is that quality comb honey may be produced at any time during the active season. As with any endeavor, there are always exceptions, but generally speaking, quality comb honey is produced consistently by experienced beekeepers, usually prior to mid-season, and from well managed, overwintered colonies. To understand this we should examine the following characteristics of quality comb honey.

■ The wax is edible. It is tender. The consumer will chew and frequently swallow it without thought.

■ It is attractive. In a given section of comb the cappings will have a consistent, preferably light appearance, indicating that the entire section was produced quickly from a single nectar flow.

- There will be no pollen or traces of brood rearing in the cells.

- There will be no propolis in or on the wax.

- It will taste good.

What are the conditions that bring about quality comb honey? Simply stated, the rapid production of comb honey that is tender, uniform in appearance, and good tasting requires strong colonies, a sustained nectar flow, hot days, preferably with warm nights, and an experienced beekeeper. All of these conditions relate primarily to the production of wax.

Although there are occasional exceptions, we will assume for this discussion that all honey tastes good. In producing comb honey, therefore, you must concentrate on the wax. Speed of production is essential. Given a good nectar flow the bees must work rapidly to keep up with it. They must build sufficient new comb to contain the ripening honey and therefore have little time to build strong, heavy comb. With less time and fewer idle bees there is less travel stain on the wax. The bees enter, do their job, and leave. In the early season there is less emphasis on the collection of propolis, which of course is a primary contributor to travel stain as well as being a component of older comb and reused wax.

A colony works 24 hours of the day when there is work to be done. With a good nectar flow in progress there is indeed plenty of work to be done, even at night. Wax secreted on warm or hot days and nights is more easily handled and manipulated. It can be and is worked into thinner cell walls and cappings, which in turn make more tender eating. If the weather continues to be warm then the wax will be uniformly thin. Alternating warm and cool weather can lead to erratic consistency of the comb, some thin and tender, some thick and tough. Continued cooler weather leads to tougher, less palatable wax.

For quality wax production to proceed, the bees must have a sense of urgency. This sense comes about from lots of nectar, lots of bees, and the space to expand. Plenty of foundation should be in place, and a large work force should be present. At the same time, emphasis should be diverted from brood rearing. A single brood chamber is in order. Best results are obtained by condensing a strong two-body colony into a single body.

The conditions that encourage the production of quality comb honey are, unfortunately, the conditions that encourage swarming. A good nectar flow, lots of bees, a sense of congestion in the brood area, the early season: all of these together can lead to swarming. This is one of the reasons that comb production is best left to the more experienced beekeeper.

The second reason is that the novice cannot bring about the conditions required for the production of quality comb. The first requisite is a strong colony early in the season. The second requisite is the experience to manage a colony of bees on the verge of swarming. A novice who has acquired an overwintered colony at the beginning of the beekeeping season may have the required bees but he is unlikely to have reached a level of competence to manage those bees appropriately. A beekeeper, experienced or not, starting with a package or a nuc, does not have a colony that is strong enough to make comb honey in the early season. By their nature and origin, packages and nucs do not mature into producing colonies until mid-summer or later, at which time the comb honey season is about over.

As will be pointed out quickly by many beekeepers, there are exceptions to all of this. There are instances where packages and nucs have produced quality comb honey. There are those novices who defy the experts and for whom everything goes exactly right. Beekeeping is rife with exceptions, but on a continuing basis you can not count on them.

Forms of Comb Honey

Four forms of comb honey are commonly produced and marketed. The differences have to do primarily with the packaging. They are: round comb, section comb, cut comb, and full frames.

■ *Round comb* is produced in 4¾" comb honey supers that contain plastic frames and fixtures. Round comb has become increasingly popular with beekeepers in recent years and to a large extent is pushing section comb off the scene. Round comb, having no corners and holding less honey than the square section comb, is more quickly and readily filled by the bees.

■ *Section comb* is also produced in 4¾" supers. This old standby in its square wooden boxes has been with us for many, many years. Though replaced by the round plastic sections in many instances, it will not completely yield its place. For many beekeepers and consumers this is the only kind of comb honey.

■ *Cut comb* is usually produced in standard shallow extracting supers, or occasionally in mid-depth supers. This form of comb honey requires a minimum of special equipment. It is popular among those beekeepers who wish to produce a small amount of comb for their own use, but it is also a very acceptable method to produce marketable honey. Standard extracting frames are used, though without wires in the frames or the foundation. The completed full comb can be cut out of the frame in four equal 4¼" sections, and each section packed in a square plastic box. This is a good starting method for first-time comb honey producers.

■ *Full frames* may be produced in any size super but are most common in the smaller sizes. The full wooden frame is marketed in a suitable box or wrapping. It is somewhat unwieldy as a consumer item and has never been as popular as section combs. With the advent of the 4¼" square plastic boxes for packaging cut comb honey, full frames have become even less common. They are most often seen as entries in honey shows.

The procedure for producing comb honey is the same no matter which of these forms is being produced. Only the configuration of the equipment varies.

Producing Comb Honey

As has been stated, comb honey is best produced by a strong colony. It should not only be strong, however, it should also be concentrated. That is, the efforts of the colony should be directed to building comb and making honey, and to some extent diverted from raising brood. A colony must always raise a certain amount of brood simply to maintain itself. However, brood rearing preoccupies a number of bees who could be out foraging. Many, perhaps most, of the very successful comb honey producers reduce to one hive body at the time they put comb honey supers in place. As much of the brood as possible is placed in that one hive body, along with the attendant house bees, and of course the field bees remain. If there is more brood in the original hive than will fit in a single hive body, give preference to the capped brood. This older brood needs a minimum of attention and will hatch quickly to bolster the work force. A queen excluder is placed over the hive body*, and comb supers, usually two or more, are put in place. Any extracting supers that may have been in place are removed. By reducing the available space for brood rearing, and by shifting the attention of the bees to honey production, we bring about the best environment for comb honey production.

Unfortunately, we have also brought about an environment that encourages swarming. To reduce this possibility it is advisable to have earlier introduced a new young queen in this hive. It is also helpful to have some bait comb in the honey supers to encourage the bees to move up there quickly. In this way we have eased two of the conditions that contribute to swarming.

The second hive body, presumably containing some brood (mostly uncapped), attendant bees, and honey and pollen, may be used to strengthen a weaker hive, or may be used as the nucleus of a new colony. Many comb honey producers set this hive body up as a temporary hive, placed back to back with the parent colony. It will remain there, living its own life until the comb honey season is over, at which time it is combined back with the parent colony.

A Method for the Novice

Actually, even with limited experience, anyone can probably produce a small amount of comb honey, and sometimes a large amount. Give it a try. This only works, of course, if your colony is ready to produce surplus honey.

---

*Not all experienced and successful comb honey producers agree on the use of queen excluders. A novice comb producer should probably use one.

When you are ready to place a super on the hive, instead of ten extracting frames use only seven or eight. Together in the middle of the super place the remaining two or three frames, containing unwired comb foundation. This should be done when there is a good nectar flow in progress.

If all goes well the colony will draw the foundation and fill those frames with honey. You must inspect frequently and remove the comb honey frames as soon as they are full and capped. Replace them with more comb foundation if the honey flow is going well. Otherwise, replace them with wired extracting foundation.

In the event that all does not go well, little is lost. If by the time the nectar flow is over those comb honey frames are not drawn out, or are not filled with honey, or are not properly capped, treat them as extracting frames. The only difference between these and regular extracting frames is that they will not be as strong. You must handle them a little more carefully when extracting. However, in each succeeding year the bees will strengthen the comb a little more until at some point you will not even recognize their origins.

A variation on this method is to use comb foundation but to cross-wire the frames, as is often done with regular brood or extracting frames. The wire can be removed at the time the honey is harvested by first cutting it on the outer side of the end bars and then withdrawing it through the holes in the end bars. This is best done by first heating the wire quickly and briefly to free it from the wax. Beekeepers have devised various methods to heat the wire, usually with a device similar to an electric wire embedder. A small current is passed quickly through each single length of wire and it is then withdrawn by pulling it out with a suitable pair of pliers.

Supply and Demand

It is interesting to listen to beekeepers discuss the apparent lack of demand for comb honey that exists today, while at the same time we hear of the continued demand for comb honey from consumers. Somehow the idea has become prevalent among many beekeepers that there is no point in producing comb honey, because no one wants it. To my view, there is a shortage of good comb honey.

At least part of the problem seems to be that few beekeepers are producing comb honey except for their own use. This being true, little is seen on the market. This absence is interpreted as a lack of demand. From my own experience I believe that the market could absorb a large quantity of comb honey over what is being produced today.

CHAPTER 10

FORAGING

▬

WE HAVE DISCUSSED FORAGING to some extent in other chapters but it is worthwhile to take a further look at this important facet of bee life. Obviously, bees must forage. Without an active field force a colony has no way to bring in the essentials of life — nectar, pollen, and water. Every worker is a forager in the last part of her life, though every one may not necessarily forage for all three of these essentials.

The normal foraging area for an individual colony is usually considered to be within a one to two mile radius. An individual bee does not start her foraging life knowing that entire range, and in fact may never learn more than a very limited part of it. Further, her range of knowledge is a function of her age. A very young field bee will know only the geography close to the hive. With age and experience she may extend her sphere of knowledge, but only if she cannot find forage near home. There is no need for her to venture far afield if there is an adequate supply of quality nectar and pollen close to the hive.

We know that bees are constant: that is, they gather nectar from only one species of plant on a given foraging trip. Beyond that, a bee may gather from that single species for the entire day, or for several days, and sometimes for her entire life as a field bee. If the nectar flow allows, an individual bee's foraging experience can be quite limited. She may spend her entire foraging life on one patch of flowers, or on one section of an orchard. As long as that particular plant gives a relatively rich return in terms of both quality and quantity of nectar our bee will stay with it. It is only after the nectar flow from that species stops, or another species is discovered by the colony to be offering a richer return, that our bee will change over.

We also know that the area covered by an individual colony may be restricted by terrain, by land use, or by plant distribution. Though the potential foraging range of a colony may be one to two miles, and individual bees may travel that distance, members of the colony may never travel in certain directions. As long as some acceptable forage is found in some directions from the hive, there is little reason for scouts or foragers to venture far into terrain that holds little promise. For instance, a colony situated on the edge of a large body of water is not likely to explore across that water if there is acceptable forage in other directions. On the other hand, if there is not acceptable forage, they will cross large bodies of water or other barren areas. Hills and valleys, wooded areas or other large acreages of non-forage plants, urban areas: all of these can affect the directions that bees will search and forage.

Bees do fly distances greater than one to two miles, sometimes traveling five miles and more. Obviously, longer distances are not desirable because fewer flights can be made in a given period of time and less of a payload results from each flight. The bee must use honey as fuel for the long outbound trip and may be forced to use some of the collected nectar on the return trip.

What does all of this mean to us in terms of practical beekeeping? There are several ways in which foraging behavior impacts hive management:

- selecting apiary sites

- determining the number of hives per site

- competition between colonies

- moving hives and the orientation of the bees

Selecting Apiary Sites

Over the years it has become more and more difficult to find good locations for bees. We have the encroachments of urban and suburban development, the reductions in farm land in so many areas, and the ever-increasing size and efficiency of mono-crop farms in other areas. These and other factors all have their effect on beekeepers, whether it be the hobbyist with one or two hives in the suburbs or the larger commercial beekeeper in a more rural area.

The average hobbyist who has one or two hives at home has little control over the larger foraging picture. The bee yard will be somewhere within the bounds of his or her property. The bees will forage in that neighborhood and beyond. For someone who is establishing an outyard, though, there are some considerations. Assuming that the goal is to make surplus honey (some beekeepers care only about pollination), is there a reasonable amount of forage within one or two miles of the proposed yard? Is the terrain such that the bees can travel to the forage freely? Are there other hives within this same foraging area or whose range overlaps it to any degree? If you must answer no to either of the first two questions or yes to the last, some further thought about the location is in order.

Number of Hives per Site

This of course also varies. It depends on the answers to the questions in the preceding paragraph. It also depends on the specific kinds of forage available. On a continuing basis an area of small diversified farming will usually support more colonies than will one with large acreages in single crops. Open rural land with its varied weeds, wildflowers and fence-row vegetation will support more colonies than will wooded areas. A large acreage of a particular crop may support many colonies while the crop is in bloom, but the area may not support any hives otherwise. Experience, either yours or that of other local beekeepers, should give some idea of how many hives might be placed in a given area. Trial and error over a period of time will show if you are right. It does take more than one year to prove a site. Once you are satisfied with the amount of forage, though, you must still pay attention to the area. Changes in land use can affect the amount of available forage very quickly.

Competition Between Colonies

Colonies do compete. Certainly this is not a conscious thing but it happens. If there are two or more hives together the foragers must spread out over a larger area in order to be successful. They will tend to explore farther and in more directions, simply because there are more foragers. This can force them to explore areas they might not otherwise have needed to go, and can possibly expose them to richer nectar sources. We can take advantage of this by keeping our hives together, within reason.

On the other hand, spreading hives through an area enhances efficiency. Bees will initially forage close to their hive, only flying farther afield when the nearby resources are no longer sufficient, or when competition from other nearby hives forces them. Grouping large numbers of hives can waste flight time since they will quickly saturate the available forage near home. They are then forced to fly greater distances to find additional forage. A pollination situation is a good example. Small groups of hives spotted around an orchard are more efficient than one large group. The small groups will compete within themselves to their own benefit, but each group will have its own somewhat exclusive foraging range within a reasonable travel distance.

Moving Hives

Many of us must occasionally move hives. A particular site may prove to be poor foraging, or perhaps a neighbor complains, or any of a number of other problems pop up. Conventional wisdom says that if you wish to move your hives to a new location that is within the foraging range of the old, you must first move them two miles or more, leave them there for three weeks, then bring them back to the new location. This keeps the foragers from becoming confused and from returning to their old home site. During the three weeks the original foragers die off and are replaced by a new crew who do not know the original terrain.

This is perfectly good advice and if you can move your hives two miles or more you should have no problems. It is not always possible or perhaps desirable, however, to move two miles. Furthermore, it may not be necessary. Look carefully at the probable foraging

LARGE AND SMALL GROUPS OF HIVES. These bees are being used commercially to pollinate cranberries. One group has six colonies, the other over 100. The smaller grouping is much more efficient.

areas for the two sites in question, the old and the new. Do these ranges in fact overlap? As stated, crop patterns, land use, or terrain may be such that from a practical standpoint there is no overlap. The normal foraging pattern for the original site may not overlap that of the new at all, or the overlap may be so small as not to warrant consideration. If this is true you may be able to eliminate that interim move, going directly from the old to the new site.

While we are thinking about moving bees, let's give some thought to small moves. It is not unusual to want to move a hive or two on your own property, sometimes just across the yard or out of the sight of an apprehensive neighbor. Again, conventional wisdom says move the hive two miles or more, wait at least three weeks, then bring it back to the new location. Again, this is sometimes not necessary.

If the terrain allows and the landmarks are not overly confusing, it is possible to move the hive a few feet a day and have the bees follow it. This might not work well if done in a wooded area, or if it involved going around corners so that the bees lose sight of the hive. It does work if the bees can easily see the hive from their old location. It is best to leave the hive in each interim location for a few days if possible. I have moved a hive ten feet at a time with no problem.

If more than one hive is involved, they should be moved together, maintaining relative positions. Otherwise drifting may occur. Given two hives sitting next to each other, if one hive is moved and the other not, almost certainly some if not all of the field bees from the moved hive will orient to the hive remaining at the old stand.

CHAPTER 11

SUMMERTIME

As we move out of the early season and into summer the beekeeper's tendency is to coast a little. The critical actions are past. New colonies have been started, swarm season is out of the way, comb honey production is well in hand, and in general, hives are producing. It is easy to sit back and forget the bees for a while.

Some coasting is all right, but be careful. Complacency can be costly. Some things to keep in mind are:

■ *Variations in honey flows over the years.* Honey flows do vary radically from year to year, especially in areas like New England where the weather can fluctuate so. This variance seems to surprise many beekeepers. It shouldn't; it's a fact of beekeeping life. But the question comes up time and again: why is this such a poor year or such a good year? Well, what are the factors? Primarily it is the weather. For plants to manufacture and secrete nectar there must be the right combination of sun, rain, and warm weather during the spring and summer. A wet spring or summer gives plenty of moisture for the plants but can suppress growth or bloom. Furthermore, though this is perhaps not a big factor, a sudden shower can dilute the nectar in some blossoms, making it uneconomical for the bees to gather. If the nectar is too diluted, that is, too low in sugar content, it may take more energy to gather it than is actually gained. Hot, dry weather has a different effect, reducing the amount of nectar available, sometimes by a lack of ground water from which to make nectar, sometimes by stunting the growth of the plant or the blossom. Some years everything comes together just right and there are bumper crops of nectar. If beekeepers are alert to what is happening and get enough supers on, that nectar is collected and becomes a record crop of honey. Other years it all gets past us. We perhaps realize what is happening just as it's all over.

■ *Supersedure.* Did the bees of your colony supersede their queen? If so, did all go well? If it was a successful supersedure, is the resulting queen of the quality that you want? Are her offspring of the quality that you want? In and of itself, supersedure is not bad. It indicates that the queen mother was failing in some way and the colony, recognizing this, replaced her. The overall effect should be positive, though. Be sure that from your point of view you have a better colony. Supersedure can take place at any time but more often it seems to take place in the late spring or early summer. Marking the queen will help you keep track of her status.

■ *Population.* During the peak of the season that has just passed, did your colony build up to an acceptable population level? When you open the hive are there bees welling up from between the frames and are the top bars covered? If the colony has not built up by midsummer, it is unlikely to do so at all on its own. A new queen at this time would be a good way to give the bees a boost, or you might add a frame or two of brood from another colony.

■ *Condition of comb.* Whether this is a new or an overwintered colony, have they drawn all of the wax foundation that you may have introduced this season, or at least is it being drawn? The closer you are to the end of the season the less likely you are to have that foundation drawn out. If it has not been drawn, this may be an indication that the colony is not up to strength or is having other problems. It is your signal to look more closely. Again, a new queen or an infusion of brood from another colony may be in order. Maybe this is the place to use that nuc that you have been keeping on hand.

■ *Seasonal nectar flows.* These are not consistent, from year to year or from area to area. Do not expect that just because a given plant gave copiously of nectar last year, it will automatically do so this year. It probably will, but there are no guarantees. A cold or wet spring, ongoing bad weather, changes in land use, or major construction in the area: any of these may have an effect. For instance, the farmer just over the hill who grew so much clover may have switched to a different crop, or may have sold out to a developer.

■ *Daily nectar flows.* These are not consistent throughout the season: that is, there is not necessarily nectar available every day of the active season, nor is the volume consistent from day to day when there is a flow. In some areas the season of active nectar flow ends in midsummer. In other areas there may be a fall flow but a mid-season dearth. Be sure colonies have ample food reserves to carry them. Especially with new colonies, be willing to feed if there is any question.

■ *Other beekeepers.* Perhaps there has been no change in the nectar flows but there is more competition for the nectar that is available. Another beekeeper just down the road might decide to expand his hobby and now has twenty or thirty hives instead of one or two. Or perhaps the hobby became more popular and there are now several new beekeepers in an area where you had previously been alone. Competition from other beekeepers can come on all of a sudden or it might build up over the years. No matter how it may come about, however, be alert to the doings of the other beekeepers whose operations may overlap your bees' foraging range.

Some specific actions may be in order during the summer. An obvious one is supering. Assuming that your management scheme involves adding supers as they are required rather than all at once in the early season, remember to keep checking. Do your hives need supers? Be sure they are needed before placing them. The hive bodies should be well used before supering. You do not want a situation wherein supers are placed on top of hive bodies that are only partially full, and the bees move on up into the supers without ever fully utilizing the brood chambers. Though queen excluders help to prevent this from happening, not all beekeepers choose to use excluders. If the bees are allowed to move up into the supers without fully utilizing the brood chambers, then in the fall there may be honey and perhaps brood in the supers but little or no honey stored in the hive bodies for winter.

Another action, if it is a part of your management system, is to remove and extract the early-season crop. Many beekeepers do not extract until the end of the fall flow. This relates to the nature of the nectar flows in your area and to your management system. Generally, early-season honey is of lighter color than that from the late season. Some beekeepers prefer not to mix the two. This may be a prejudice against dark honey on the part of the beekeeper, or it may be dictated by market conditions. Many consumers prefer the lighter grades. Of course, many others, beekeepers and consumers alike, prefer the darker honeys.

Now, or at any part of the season, is a good time for some routine housekeeping chores. Look at individual frames as you work the hive. Are any of these frames candidates for replacement in the following spring. If so, mark them now so that they are easily identifiable later, and then start working them towards the outer wall of the hive body. If any of these are brood frames in the center area of the hive, move them one position each time you have the hive open for routine inspection, so that by the end of the season they are next to the wall of the hive body. Place them in the bottom hive body, if possible, since that one will normally be empty in the spring and the frames can be removed with a minimum of disruption.

Also identify and mark any other pieces of hive equipment that are candidates for repainting or reconditioning. Wooden hive parts that are properly cared for have the potential for a very long life. If routine maintenance is ignored, they can rot out very quickly. There is nothing more disheartening than to have to move a hive in a hurry and to discover that the bottom board has rotted away to the point of disintegration. Winter is a good time for reconditioning equipment, spring is a good time for the actual replacement, but identifying those parts requiring attention is a year-round activity.

CHAPTER 12

LATE-SEASON MANAGEMENT

W<small>HAT IS THE LATE SEASON</small>? To me it is a loosely defined period that may begin as early as the first of August and extends into the preparation for winter. During this time we begin to think of the winter to come. In some parts of the country this is perfectly logical: the main nectar flows are over and the crop is made for the season. But for those who live in areas with a good fall nectar flow, as in New England with the goldenrod and asters, August may seem much too early to think about winter. It is not too early. This is the time to think of supering for the fall flow, of requeening, and of combining weak colonies. Perhaps you do not actually do all of these things in early August, but you do think of them and begin to observe and plan.

Supering for the Fall Flow
If you do not have a fall flow in your area there is no need for any further supering, and if you have a continuous flow throughout the active season, your hives have been supered up right along. There are some areas though where there is an early season flow and a late season flow. Some beekeepers in these areas extract between flows. If

that is the case, then be sure that you have replaced the supers before that fall flow begins.

Requeening

Requeening can be done at any time during the active season, though there are times that are better than others. Emergency requeening of course is done when the emergency arises. Routine requeening though is usually done in the spring and fall. However, for requeening purposes fall should be defined as late summer. Actual fall, that is, late September and beyond may be too late. If requeening is done in the late summer instead of in the fall there is then ample time for the new queen to be accepted, settle in, and lay plenty of eggs before winter. The resulting new crop of young bees is important to successful wintering, and a new queen will normally produce a bumper crop of those young bees. Conversely, if requeening is done late and is unsuccessful, there may not be time to try again.

Late-Season Requeening. At any given time, a queen is laying eggs at a rate that is a function of several things; the time of year, the needs of the colony, and her age and capability. Once the season has passed its peak, there will be a natural slowdown in egg laying, even if the queen is young and vigorous. If there are no extraordinary events such as disease, pesticide kill or predation, the needs of the colony will allow for this natural slowdown. And even if a queen is aging, any reduced capability may be masked by the seasonal reduction in rate. However, if a colony is requeened with a new young queen, she will normally lay at a higher rate than the queen that she is replacing even if the queen being replaced was relatively young and vigorous. In other words, the factors governing the colony's rate of egg laying are temporarily overridden by the introduction of a new queen.

This is a major reason for late-season requeening. The new queen will lay at a higher rate. Consequently, the colony will have a higher proportion of young bees who are better able to withstand the rigors of winter and who will be available in the spring to help get the colony off to a good start in the new season. The presence of this still young queen in the spring will also be a key factor in controlling swarming.

Combining Weak Colonies

As has been discussed, a colony that has not built up by late summer is not likely to build up at all. If it has been a problem colony and is still weak in late summer something is basically wrong. A colony may temporarily have less than normal population but still be in basic good health, with the potential to make it through an average winter. Another colony, though, may be inherently weak. It required attention earlier that it did not receive. The appropriate attention in late season tends to be drastic. Combine such weak colonies with other, stronger colonies. Take your winter losses now, on your own terms.

Some beekeepers resist such an action. They do not want to reduce the number of hives they own, or they simply procrastinate, hoping that the situation will improve. Almost certainly, it will not.

Once the decision has been made to combine, then the earlier the action is taken the better. The two colonies involved must have time to come together, merge their brood rearing, reorganize food stores, and in general get ready for winter. When combining hives, it is usually best for you the beekeeper to determine which is the better queen, and then dispose of the other. Many beekeepers are of the opinion that when two queens end up in the same colony they will meet and fight, and the best one will win. This is not likely to be the case. Usually it is the workers, encountering a strange queen in their hive, who kill off one of them.

Selecting Colonies to Merge. If you own only two colonies, your choice of merge mates is made for you. If you have several colonies, you have flexibility. Merge your weak colony with a strong colony. Do not merge two weak colonies. The arithmetic here is simple. Two weak colonies combined equal one larger weak colony. This has been demonstrated many times. If you have more than one weak colony, combine each one with a strong colony. Three weak colonies combined also equals one larger weak colony.

If you combine weak with strong, you should end the winter with strong colonies that are candidates for being split. Ideally you can split in the spring and start the new season with as many colonies as you had before combining in the fall. Your only expense will be for a new queen to put in the new colony that you split off in the spring.

CHAPTER 13

TAKING THE CROP

HARVESTING THE CROP IS A CHORE that looms large for many beekeepers. It often seems to be such an overwhelming task that it simply does not get done the first year or two of keeping bees. Novices sometimes leave honey supers in place the first winter rather than face the problem. This of course can lead to a further problem in the spring when they may find that the colony has moved up into the honey supers over the winter and is happily raising brood there.

The question to be addressed then is — how am I going to get those supers full of bees off the hive? The literature talks blithely of brushing, of bee escapes, of chemicals, but often leaves out the how, the why, the when. There are many specific questions that you might ask. Some of the most common are:

■ When do you take off honey? That is, do you wait till the end of the season, or do you start taking it as soon as there is a full super?

■ How much do you take off?

■ Is it all right to leave honey supers on for the winter?

■ Should you remove the queen excluder for the winter?

- What do you do with unripe honey in the supers?

- What do you do if you have some honey in the supers but not enough in the hive bodies for wintering?

- What do you do with wet supers after extracting? Is it all right to put them outside for the bees to rob?

The answers to some of these questions are interrelated and some of them have been discussed in other parts of the book, but let us consider them again in the context of taking the crop.

When do you take off honey? There are a couple of answers to this question. It depends in part on the nectar flows in your area, and in part on your preferences regarding honey. In some areas the main nectar flow is over by mid-summer, with nothing significant beyond that. In other areas there is a somewhat continuous nectar flow through most of the season. In still other areas, a main nectar flow occurs in the early summer, followed by another significant flow in the late season, with the honey from the early flow usually being lighter and milder. Many consumers have a conditioned preference for the lighter honey. Many beekeepers are prejudiced against the darker and sometimes stronger-tasting honey from the late season, considering it good only for winter stores or for sale as bakery-grade honey. This is an unfortunate attitude. Very few honeys do not taste good, but they do come in many flavors. Too many people expect all honey to taste the same and too often they expect it to be that bland blend of honey frequently found in the supermarket. Anything different is suspect.

A first question to be answered is — what is the nature of the nectar flows in your area? A second question: do you have preferences or prejudices towards any of the seasonal varieties? Your choices of action which follow are:

- If you have more or less continuous nectar flows throughout the season you may choose to wait and take off the honey at the end, or you may take it at intervals through the season. If you are able to separate varietal flows, this latter may be desirable. If you have large crops, you may wish to spread the work over more than one extraction and you may wish to minimize the number of supers you own by reusing them during the season.

■ If you have a single early flow, you will normally take off honey once, at the end of your season.

■ If you have both an early and a late flow, there are two approaches. Take off honey after each flow, or wait and take it only after the final flow. The former approach will give you two separate crops, probably of different color and taste, while the latter will give a blend. The former will also be more laborious while the latter allows you to own fewer supers.

A basic rule of thumb is: do not take the honey until you are sure that the colony does not need it for winter. This can be overridden of course if you are prepared to feed to compensate for taking those potential winter stores, but it is questionable if this should be your deliberate approach.

How much do you take off? Simply stated, you take off only that which is surplus to the needs of the colony. To establish what these needs are, ask yourself these questions. Is this the end of the main nectar flow? Are there any minor flows to come? How long is it until winter begins? How much will the colony consume between now and then? How much will they need to make it through the winter? Though there may be honey in the supers, is there also the requisite amount in the hive bodies? Your goal is to come to the end of the season with as much honey or stores in the hive bodies as is required for overwintering in your area.

Precise amounts are difficult to state here. Requirements vary throughout the country. The practices of your area, modified by your experience, are your guide. This is further discussed in Chapter 14.

Is it all right to leave honey supers on for the winter? Though there are exceptions, generally it is best not to leave supers on the hive during winter. There are two potential problems. Two deep hive bodies will suffice for winter quarters and honey storage in just about any part of the country. If the reason for leaving a super on is because otherwise there may not be enough stores for winter, then this means that the hive bodies have not been utilized efficiently. They are not full. Even though leaving a super may give them enough food reserves, the food may not be well distributed in the hive. It may be possible for the bees to become isolated from the honey if it is not efficiently stored, especially in very cold weather when movement in the hive is restricted.

A second problem is that in the spring the colony is most likely to have worked its way up and is living and raising brood in the super. You then face the problem of clearing the super so that it can be returned to its normal function. This is not a major problem but it is at least a nuisance. The super can most readily be cleared by placing it under the bottom hive body in the spring. The colony will normally move up out of it to the hive body above as the brood matures. The super of course will suffer some minor wear, tear, and darkening from being used to raise brood.

Should you remove the queen excluder for the winter? There are occasions when you may decide to leave a full super on the hive for the winter. Even so, the queen excluder should be removed. Otherwise the colony, or at least the bulk of the colony, may move up through the excluder as the season progresses, with the danger of leaving the queen behind. At worst she could be isolated and die; at minimum she would not be able to develop an efficient brood nest as the new year begins.

What do you do with unripe honey in the supers? First, be sure that the honey is in fact unripe, and not simply uncapped. It is not uncommon in the late season for the bees to leave a larger than normal amount of honey uncapped. To make this determination, look at the honey: cells of unripe honey usually will be little more than one half full. Hold the frame horizontally and shake it: unripe honey will drip out. Stick your finger in and get a taste: unripe honey will taste "green."

If you determine that you are dealing with unripe honey, feed it back to the bees. They will ripen it before winter. Feeding back is most easily done by putting the super over the inner cover in the late season, first scratching any cappings that do exist. Your goal is to do a little damage (only a little) to the comb and cappings so that the honey will ooze and run a bit. The bees will be prompted to go up through the hole in the inner cover and remove the honey, bringing it down and storing it in the hive bodies. This must be done in the late season, as the nectar flow is winding down. Done too early, the bees will not move the honey. They may ignore it completely or they may repair any damage that you have done and continue to use the super, even though it is above the inner cover.

What do you do if you have some honey in the supers but not enough in the hive bodies for wintering? The same technique applies here as for unripe honey. Scratch the cappings, put the supers over the inner cover, and let the bees move it. If there is more honey in the supers than the bees need, you can extract it first, and then feed back what they need in a conventional feeder.

What do you do with wet supers after extracting? Is it all right to put them outside for the bees to rob? You have several choices for dealing with wet supers. The first is to store them wet. The residual honey will probably granulate over winter but the bees seem to have no problem dealing with it at the start of the next season. However, on warm days during the winter and early spring these wet supers will be attractive to the bees and you will have to cope with that unless you put them under bee-tight cover. You can eliminate the problem by having the bees clean out the supers right after you extract.

There are two ways to do this. One is to put the supers back on the hives and let each colony clean up its own. The supers are placed back on the hives, over the inner cover, and in one or two days the residual honey will be cleaned out and stored below. The other approach is simply to set the supers out somewhere and let the bees have a big clean-up party.

Conventional wisdom says to never allow bees to rob. There are at least three reasons. First, robbing begets robbing. If bees are allowed to rob supers, then when all of the honey has been cleaned up they may very well search for more. If the first thing they come across is a weak or relatively weak hive, you have trouble. Second, bees often become frenzied when robbing, damaging the comb that they are stripping dry and sometimes fighting and killing each other. Third, robbing is a possible way to spread disease.

I often put out stacks of supers to be robbed. However, I do have a method. My method eliminates two of the potential problems and ignores the third. First, I only do this if there are plenty of supers, at least two for every hive that may participate in the robbing. The supers are piled in staggered stacks so that there are plenty of places for the bees to get in. There is an entry to every individual super so the bees do not have to compete to squeeze through an entrance. With plenty of supers and plenty of entrances, fighting and competition is largely eliminated. The supers are stacked well away from the bee yard, at least 100 yards, so that the robbing behavior is not encouraged at home. The supers are left in the stacks for several days after they have been cleaned out so that the bees become completely aware that they have been emptied. Storing away the supers too soon, before the bees accept that the source has gone dry, could lead to further searching and the possibility of hives being robbed.

One further consideration, and an important one, is that I am in an area isolated from other beekeepers. There is no danger that I will be interfering with someone else's bees, causing them to initiate robbing.

Methods for Removing Supers

In order to take the crop, the supers must be removed from the hive and the bees removed from the supers. There are several methods for doing this. Let's deal with taking the supers from the hive first.

In most hives there is a certain amount of burr comb built up between supers, or between the top super and the inner cover, or between the top hive body and the lower super. Lifting the cover or lifting the individual supers usually breaks this comb and causes honey to ooze and run. Honey is then spread around as you manipulate equipment and the odor of honey is in the air. Robbing may be triggered if you continue working with the hive open.

A way to relieve this problem is to go to the hive the day before you plan to remove the supers and at that time break up the burr comb. Do this by quickly removing the cover and each super, immediately replacing them in a different order or orientation. Close the hive and leave. By moving each super you will have broken up the burr comb but you will have left the oozing and running honey inside the hive where the bees will deal with it, most likely storing it down below in the supers or hive bodies where it belongs. When you return the next day you will have a cleaner situation to deal with as you remove the supers.

Even doing this, there is still the potential for robbing when you open hives to remove the supers, or even at other times when you are routinely working your hives. You can ease this problem by taking the covers off of all of your hives when you first come to the bee yard. Each hive then is put on the defensive. They are more concerned with protecting their own stores than they are with robbing other colonies.

There are at least four generally accepted methods for removing bees from the supers as the supers are taken from the hive. These are: brushing, an escape board, a fume board, or a blower.

Brushing

Brushing is easy, inexpensive, and relatively quick. If many supers are involved, though, it becomes tedious and it can be very difficult to keep the bees from returning to the frames even as they are being brushed. To do this, take the supers, one at a time, a few feet from the hive. Preferably, have an empty super or hive body ready to receive the frames as they are brushed clean. Place the empty on a solid bee-tight base and throw a damp cloth over the top to act as a temporary cover. An old towel works well. The dampness helps to hold it in place and allows you to manipulate it more easily as you flip it open to place each frame after brushing. The bees will be trying very hard to get in with those frames.

The actual brushing is straightforward. Hold the frame near the hive entrance, take your bee brush and brush gently but firmly. Work as quickly as possible. When the frame is clean of bees put it in the empty super under the cloth cover. Use a proper bee brush. A handful of tall grass is often recommended for brushing bees. This may work for a quick brushing of one or two frames as you are routinely working a hive. It is makeshift if you are brushing many frames such as when you take off honey.

Escape Board

The escape board is available in several forms. The most basic is the inner cover with a Porter bee escape set into the hole in the center. Other designs are available but each serves the same basic purpose, that of providing a one-way exit for the bees, out of the supers and into the hive body below. There must be no alternate entrances into the super from above or below.

In normal use, the escape board is placed under the supers and left there for about twenty-four hours. In theory, after that time all of the bees will have gone down through the bee escape, leaving the super empty and ready for removal. In practice, however, it does not always work. If any brood happens to be in the honey supers the bees will usually not vacate, or if the bee escape is not working properly they may not be able to exit. Even when everything goes well, a few bees may not exit and it will be necessary to do a little brushing.

Fume board

The fume board is an absorbent board that is used in conjunction with a foul-smelling chemical. The bees are driven from the super by the unpleasant odor. The fume board is placed on the hive directly over the super to be emptied. If all goes well, the super is emptied in a matter of minutes. The weather is important: a fume board works well only on a warm, sunny day. Even then it will drive the bees from only one super at a time, and rarely works well with a full deep super.

Several different chemicals have been available over the years for use with a fume board. Do not experiment. Use only one of those that have been approved for the purpose, as are available from your beekeeping supply dealer. Take no chances on contaminating honey.

These chemicals are truly foul-smelling. Do not store them in the house, and whatever you do, do not spill any on your clothing. If you do, even your best friends will shun you.

Blower

A moderately strong current of air is a surprisingly effective way to remove bees from a super. There are blowers listed specifically for the purpose in bee supply catalogs, but I have found that a simple leaf

FUME BOARD. A foul smelling chemical spread on the cloth lining of the fume board will drive the bees out of the super in a matter of minutes.

blower works well. Most vacuum cleaners (reversed) do not give a strong enough blast.

To do the job it is helpful to have a stand of some sort on which to place the super. I use an empty hive body, but a stand with legs is a little more stable. Place the super to be emptied on its end on the stand, so that the frames are vertical. When the current of air is blown between the top bars of the frames, the bees will be blown out between the bottom bars and towards the hive entrance. It sometimes takes a little maneuvering with the nozzle to get all of the bees. They

BLOWER. This type of blower is commonly used for blowing fallen leaves from the lawn. It works well to clear bees from supers.

will hide behind bridge comb and often cluster on the bottom of the super where the air current is weaker. You may have to get the last few with the brush.

The cleaner that you are able to keep your supers of bridge and burr comb, the more effective this method will be.

CHAPTER 14

FALL PREPARATION

▬

FALL MANAGEMENT AND PREPARATION FOR WINTER are the culmination of the season's activities. The long season is approaching its end and both you and the bees are about to enjoy a rest. There are things to be done, though, and they are important to the long-term success of your colonies. Do not shirk any of them.

There are four requirements to be met for a successful winter. Since you have been a faithful reader and have been following along with the actions and suggestions of earlier chapters, you are well on your way to having your hives prepared. The four things your colonies require are: a healthy and vigorous queen; a strong, healthy population; sufficient food reserves; and a good hive location with protection from the elements.

The queen, colony population, and current colony health should at this time be givens. Your actions throughout the season have led you to the point where you have a vigorous and healthy queen, an adequate population, and good health in the general population. This leaves three areas requiring your attention: food reserves, medication, and hive protection. We will consider these in order.

Food Reserves
By food here we mean both honey and pollen. There must be enough of both in the hive to take the colony through until the season is

solidly underway in the following spring. Though both nectar and pollen are available as early as February in some areas, in much of the country this is not a serious beginning to the season. The winter food reserves should be enough to support the normal life and growth of the colony at least into April, and in some northern areas, even later.

The actual amount of stores will vary with your area and local conditions. If you are not certain how much is required for your area, ask around. In much of New England and New York, for instance, 65 to 70 pounds of honey is the accepted amount of reserves for a comfortable winter. Assuming five pounds of honey per deep frame, this means the equivalent of thirteen to fourteen frames full of honey on hand when the hive is buttoned up for winter.

Though the adults of a colony may survive the winter with no pollen in the hive, pollen is essential to brood rearing. It must be present for normal colony growth in the spring. If there is no pollen, the colony will not attempt to raise brood. The equivalent of three to four frames full of pollen is usually enough. Special notice should be taken of the amount of pollen in the hive in the fall. The bees will normally store a good supply in the late season, an amount that should get them through until spring. Occasionally, though, a colony will be short. There is little that you can do about this in the fall, other than be aware of it. Though pollen or pollen substitute may be fed successfully in the spring, bees do not accept and store it in the fall. You must store away the fact of the shortage in your memory or, better, in your journal, and be prepared to feed in the late winter or early spring.

Medication

Diseases, mites, and medication are a fact of beekeeping life. A detailed discussion of these, however, is not within the scope of this book. Refer to Appendix A for further comment on this. We will assume here that your colonies are to be medicated for foulbrood and for nosema. Terramycin may be fed in syrup, and Fumidil-B must be fed in syrup. Following the manufacturer's recommendations, the amount of syrup to be fed is at least two gallons. The recommended mixture for fall feeding is two parts sugar to one part water. Two gallons of 2:1 syrup yield about 10 pounds of stores in the hive.

A question to be answered here — does the colony have space to store the syrup you are about to feed? It is surprising how much syrup a colony can tuck away in a hive that appears to be full. However, if it is really full you must make space. The simplest way to do this is to remove frames of honey and replace them with frames of empty, drawn comb. The honey that is removed may be placed in a hive that does not have adequate reserves, or stored in the freezer as a reserve against emergencies, or it may be extracted for your own use if you are confident that all is well. Some beekeepers choose a slightly different approach. They extract it, thereby preventing granulation in the comb, then store the extracted honey away for possible future use as feed. If it is not needed for feed it is ready for your own use.

Hive Protection

Colonies need several kinds of protection in the winter; from wind, from excess moisture and carbon dioxide, and from mice. Each of these protections can be provided easily. The degree of protection, though, is variable, depending upon specific location, general climate for the area, and the beekeeper's attitude or philosophy of beekeeping.

Beekeepers generally choose and set up new bee yards in the spring or summer, and often give no thought to what the winter conditions may be at that location. Those winter conditions can be difficult. Wind, drifting snow, spring melt and runoff, all of these can make life in the hive a bit harsher and more difficult. The first step then is to consider year-round conditions when setting up a bee yard. The second step is to compensate for those adverse conditions that cannot be changed. If there is in fact too much wind, a windbreak is in order. On a temporary basis, snow fencing or a couple of bales of hay piled on the windward side may suffice. On a more permanent basis, shrubbery or a section of permanent fence are in order.

Mice appreciate the relative warmth and protection from the elements that are offered by a bee hive. It is a favorite wintering spot. At one time or another every beekeeper who does not take steps to prevent it will probably find a mouse living in one or more of his or her hives. Such prevention is easily given: an entrance reducer or mouse guard made of metal that a mouse cannot chew through will do the trick. There are such devices commercially available, or they

MOUSE GUARD IN PLACE. The bees have reasonable access through the ⅜" holes in the mouse guard, but no mice can enter. It is helpful to remove any pileup of dead bees from behind the mouse guard once or twice during the winter.

can be made up quickly from wire mesh of a sufficiently small size. A ⅜" mesh works nicely. Many beekeepers depend on a standard wooden entrance remover to keep out mice but any long-term success with this almost certainly means that no mouse was interested in entering the hive. Any self-respecting mouse can enlarge the opening in a wooden reducer very quickly. Be sure that the mouse guard is in place early enough, say late September. Mice do not procrastinate the way humans do. They recognize the onset of winter and get ready in plenty of time. Before placing a mouse guard, it is prudent to check that a mouse has not already moved in. You do not want to trap one in there.

Cold as such does not harm a healthy, well-fed colony that is housed in a reasonably secure and protected hive. Bees do not need to heat the inside of the hive, nor do they attempt to do so. They keep themselves warm and in so doing they heat the cluster and the brood nest. Of course, heat does dissipate from the cluster into the atmosphere of the hive. The hive interior does not get warm however. This is natural and accepted. Moisture and carbon dioxide, which are a natural product of metabolism and respiration, are given off by the cluster and can accumulate in the hive in unacceptable quantities if care is not taken to prevent it. Top ventilation to take care of the moisture and the carbon dioxide is imperative, but it must be balanced with the need for some stability in the hive. A constant cold draft is neither desirable or necessary.

The degree of protection given to a hive usually reflects three things: the climate, the practices of the area, and the beekeeper's own outlook or philosophy. The type of protection can range from none to complete insulation and wrapping, somewhat tempered by the number of hives involved. That is, the hobbyist with one or two hives is often willing to spend more time and effort per hive than is a larger operator with many hives.

The method that I favor is a popular one and is adequate for a substantial part of the U.S. where any kind of protection is required. It includes an adjustment to the inner cover, an insulating board, and a metal mouse guard. I normally use a slatted rack and this becomes an integral part of my wintering configuration.

Beginning at the bottom, each hive has a slatted rack. Slatted racks have a wide board in the front which serves as a buffer. The actual entrance is effectively moved back into the hive about four inches. Because of this I do not consider it necessary to use a conventional entrance reducer to restrict the area of exposure. Instead, a metal mouse guard is fastened over the regular entrance as shown. The fact of a larger area of opening does not automatically mean a rush of cold air into the hive. There can only be as much air flow as the upper entrance allows. Moving to the top of the hive, the inner cover is inverted so that the rim is down and the flush surface is up

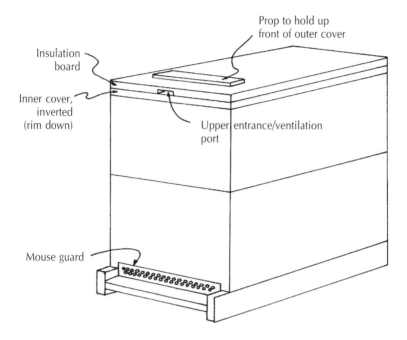

FIGURE 14-1. Hive prepared for winter.

(scraped clean of wax and propolis). An insulating board is layed flat on the inner cover and a ⅜" slat is then layed across the front of the insulating board. The outer cover is then put in place.

The inner cover has been turned upside down from its normal summer position, eliminating any air space between the insulating board and the inner cover. This in turn eliminates the possibility of condensation on the underside of the inner cover, as could happen if there were cold air immediately above it.

The insulating board has a groove on its underside that is about ¼" deep by ¾" wide and runs from the center of the board to the front edge. This groove serves two purposes. It allows a reasonable flow of air through the hive, and thus serves as a means of ventilation. It also provides an upper entrance from the hive. The board itself serves as a wick, absorbing moisture from the hive, which is in turn evaporated from the top and edges of the board, out into the atmosphere. This is

INSULATION BOARD BEING PLACED. A one-half inch thick insulation board is being placed over the inverted inner cover. The bees have an upper entrance and ventilation port through the hole in the inner cover and out through the groove in the insulation board.

facilitated by the air flow over the top of the insulating board that is allowed by the board being propped up slightly by the slat. The outer cover when put in place rests on the slat. The outer cover now rides a little higher on the hive since there is an extra layer in place beneath it. This almost guarantees that the outer cover will blow off the hive at some point during the winter unless a weight is placed on it. A pair of bricks or something of similar weight will usually suffice to hold it in place.

Wrapping Hives

The arrangement that I have described has worked well for me in plant hardiness zone 5. Some beekeepers, though, prefer to wrap their hives with tar paper or with specially made hive-wrapping materials. Farther north, in zones 3 or 4 for instance, wrapping may very well be a necessity. Again, check local practices.

If you do choose to wrap, remember to allow for upper ventilation. Unless specific care is taken, wrapping can result in an air-tight cover over the hive which will make wintering very difficult for the colony.

CHAPTER 15

EQUIPMENT

$\mathbf{A}$N ENTIRE DIMENSION OF BEEKEEPING IS THE EQUIPMENT. This equipment, for the most part, has evolved to meet the needs of the bees first, the beekeeper second. This must be or the bees will not accept the hives that we give them. If the hive is not accept-able, the bees will abscond. Of course, over the years much equipment has appeared that has not truly met the needs of the bees, or the beekeeper for that matter. Such equipment does not survive in the marketplace for long. Beekeepers do reinvent the wheel regularly, though. One of the best defenses against doing this is to read some of the old literature — books, magazines, catalogs — whatever you can put your hands on. If for no other reason, read these because the history of beekeeping is fascinating.

Anyone reading this book is assumed to have been keeping bees for a year or two at least and therefore has an understanding of most standard equipment. We will not belabor the obvious. However, there is supplementary information that is worth discussing.

Bottom Boards
Most bottom boards shown in catalogs are of the so-called "reversible"

style. These are made so that either side may be turned up. One side has a depth of ¾" or ⅞". The reverse is usually ⅜". This style dates back to the time when many beekeepers routinely reversed their bottom boards for the winter, putting the shallower side up. This helped reduce the inside volume of the hive and in theory made the hive easier for the bees to heat in the winter. With the increased understanding of bees and their wintering habits and requirements, we now know that it is not necessary to reverse bottom boards. The style persists however, at least partly because it is a sturdy design. A few beekeepers always use reversible bottoms with the shallow side up. This is not necessarily wrong but it does cut down on cluster space and can reduce ventilation in the hive. Deep side up seems the better way.

Slatted Rack

The slatted rack has a long history. It first appeared in the late 1800's, but then disappeared from the larger scene until a few years ago. Its present design is slightly different from the original but its purpose and use are unchanged. It is a device that is placed on the bottom board, under the lower hive body, and becomes a permanent part of the hive structure. Its purpose is at least twofold. It gives additional cluster space in the hive, helping to reduce the sense of congestion in a busy hive and improving the ventilation. Because of the wide board built into the front end of the slatted rack, it also provides a buffer between the entrance and the lower brood frames, which helps to overcome the reluctance of some queens to lay in brood comb that is close to the entrance. A further benefit that I have seen is that it also seems to help the bees to keep a cleaner hive. My experience with this has been that hives with slatted racks have a much cleaner bottom board in the spring. Why this should be is not clear. One would think that the rack might have the opposite effect, since it adds a layer of slats and slots to the hive, but for me that has not been the case.

Some beekeeping authorities have expressed doubt over the value of the slatted rack. Without question, it adds to the cost of a hive and is an additional piece of equipment to maintain and manipulate. Both of these considerations are very important to large operators, less so perhaps to hobbyists and others with small holdings. My own experience has been positive.

Frame Rests

Hive bodies and supers are made to contain frames that hang from rabbets cut into the end boards of these boxes. An industry standard has been that the rabbet for 9⅝" boxes is ¾" deep and for the smaller boxes, ⅝". Occasionally we encounter a box with ⅞" rabbets. Metal frame rests are available to be fastened into these boxes to reinforce, protect, and preserve the wooden rabbet from excess wear. Because the bees tend to propolize the frame ends and frame rests heavily, they are subject to a great deal of scraping and stress as the beekeeper works and cleans the hive. Metal frame rests can make it easier to scrape these areas clean.

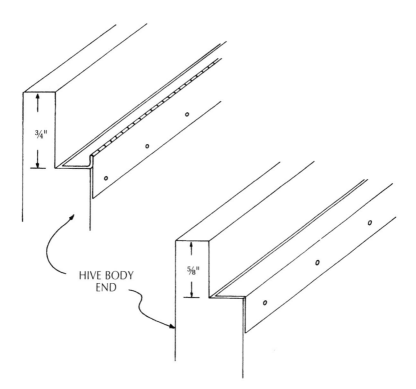

FIGURE 15-1. Two types of metal frame rest.

Two different "sizes" of metal frame rest are available. One size is for use with ⅝" rabbets, the other with ¾" rabbets. The reason for this distinction is to maintain uniformity in the vertical spacing of frames in all the boxes throughout the hive. That is, in all boxes, the tops of the frame bars should be equidistant from the tops of the boxes. This helps to standardize vertical bee space throughout the hive.

So, if you have a box with a ¾" rabbet you would normally use the style of rest that raises the frame about ⅛" above the normal top surface of the rabbet. If the rabbet is ⅝" deep, then the right-angle frame rest is in order and the frames will not be raised. This way, from box to box, all frames are at the same relative height and standard spacing is maintained throughout the hive.

Inner Covers

Inner covers are normally used with telescoping outer covers, but not with the so-called migratory covers. Like bottom boards, inner covers are reversible, with one side having a ⅜" rim and the other side flush. The normal position is with the rim up. If the rim is placed down in the active season, it creates excess space between the under surface of the cover and the tops of the underlying frames. In a vigorous colony this is an invitation to the bees to build burr comb, effectively gluing down the inner cover, to the frustration of the beekeeper.

A few beekeepers use an outer telescoping cover without an inner cover. The frames then are often fastened to the cover with burr comb or propolis and the cover is more difficult to remove. The bees are unnecessarily disturbed each time the beekeeper opens the hive. The inner cover also gives a measure of insulation in both winter and summer.

Porter Bee Escape

The Porter bee escape is a source of frustration for many beekeepers. Its cost is low, its size is handy, but in operation it does not always work. One of the big reasons for it's not working is entirely within the control of the beekeeper: maintenance. The Porter bee escape must be kept clean and adjusted. Through heavy use the opening between the spring tips may become too wide. Propolis or other debris can build

This inner cover, shown with the rim up, was on a hive during the active season with the rim down. Burr comb filled much of the space between the cover and the frame bars below.

up inside, jamming the springs. Dead bees may block the passage. The original Porter escapes were made of metal that tended to rust with time. They then became difficult if not impossible to disassemble and clean. Many beekeepers did not realize that they could be disassembled. In recent years plastic has been substituted for metal in the body of the escape and these escapes are much easier to maintain. The one or two minutes to disassemble, clean, adjust, and reassemble are well spent.

To do this, slide the two sections apart longitudinally, remove any debris or propolis, check that the springs will move apart freely, adjust if necessary and reassemble. About ⅛" between the tips of each pair of springs is good.

Escape Boards

There are several designs of escape board that are intended to work with the Porter bee escape. The most common is the inner cover with its pre-cut slot. If you use the inner cover from the hive being worked, the only extra piece of equipment required is the Porter escape. Several other designs, both homemade and commercially available, are modifications of the inner cover. Most of them have some amount of screening in them so that there is air flow and communication through the board. The bees seem more willing to go through an escape board when they know what is below, and these screened boards generally work better than do the solid wood inner covers.

Nuc boxes

Nucleus colonies have been discussed in the context of making increase. Nucs have other uses, though, and a nuc box is a handy thing to have around. For instance:

■ *Simple queen rearing.* A frame that contains one or two queen cells may be taken from a normal hive and put into a nuc box along with two or three frames of brood, bees and honey. The queen will emerge, mate and return to head up this nucleus colony, and will be available for whatever use you choose to make of her.

■ *Swarm control.* Removing two or three frames of brood from a hive that shows some early signs of swarming and placing them in a nuc immediately relieves some of the pressure of crowding in that hive.

■ *A resource.* Once established, a nuc may be kept on hand for all or part of the season as a source of brood or a queen. In an emergency the queen may be transferred into a queenless hive, or frames of brood may be used to bolster a weak hive. Afterward the nuc can be left to raise another queen and the action may be repeated as often as necessary and as time allows. If the nuc is intact at the end of the active season it may be used for routine requeening, or to bolster a weaker hive. In some areas it may even be overwintered, usually over a stronger colony using a screen board.

Nuc boxes can be purchased but they show up in very few beekeeping equipment catalogs. Most beekeepers probably make their own. Since they are not normally used year round and do not receive the wear and tear and stress of regular equipment, they can be made a little more simply than other hive parts. Most nuc boxes are made to accommodate from three to five standard deep frames, and often are made in the form of a box with the bottom permanently in place. The cover is usually the migratory style, that is, a flat board, used with no inner cover.

An important consideration in making a nuc box is the depth. Be sure there is plenty of room at the bottom. This allows you to transfer in a frame with queen cells on the bottom bar without fear of crushing those queen cells on the bottom board of the nuc.

A colony housed in a nuc box is often relatively weak, and is a prime candidate to be fed. It is also, being small and weak, a prime candidate for being robbed. If you are making your own nuc box, consider building a port in the rear of the box where a Boardman feeder can be plugged in. A Boardman feeder is often an invitation to robbers, located as it is at the entrance. It is in effect on the outside of the hive. When the feeder is placed at the rear, away from the entrance, any potential robbers will be forced to enter and travel through the length of the hive to get at the syrup. When no feeder is in place the port can be closed.

Frame Lifter Hive Tool

The frame lifter has been available in this country for several years now. It is such a great tool that I am always surprised to come across beekeepers who do not know about it. For me the advantage of using a frame lifter is in the ease with which a frame may be removed from a hive body or super. The frame lifter, with its slender, curved end, can be slipped under the end of a frame top bar. Using the adjacent frame as a fulcrum the frame lifter is rotated sideways, exerting vertical pressure only, upward on the frame being removed, downward on the fulcrum. No side pressure is exerted on the hive body or frames and it is not necessary to push the point of a tool into the frame in order to pry it out. The wear and tear on woodenware is greatly reduced.

Buying used equipment

It is always a temptation to buy used beekeeping equipment, and some great bargains quite often can be found, but this practice can lead to problems for the unwary or uninformed. Generally speaking,

FRAME LIFTER IN USE. Simple leverage works to lift the frames with no damage to the equipment, even in a crowded hive. To keep the top bars from being pried off the frames, cross nailing is important.

the buying and selling of used woodenware (without bees) by novices should be discouraged because of the very real possibility of transmitting disease, primarily American foulbrood (AFB). More experienced beekeepers, in theory at least, should know how to look after their own interests, but disease is a problem for everyone. AFB spores have been known to survive for more than thirty years, and they can withstand extremes of heat or cold. It is almost impossible to inspect equipment that has been stored away for some time and know if it once housed diseased bees, especially if wax moths have been at work.

If you do run across a deal in used equipment that you just cannot pass up, there are some precautions that you can take. Know what these precautions are and factor their cost into any deal that you may make for this equipment. One such precaution would be to begin medication of any colony you may house in this equipment, assuming you do not already medicate routinely. Other precautions include scorching interior surfaces of the woodenware, melting or burning the comb, boiling empty frames in lye water, and destroying any questionable woodenware by burning.

Be sure to do something. It can be devastating to destroy a colony and to burn the hive equipment, but this is often the only recourse in a serious case of AFB.

The possibility of disease being transmitted through other equipment should not be ignored. Most used equipment other than woodenware can be cleaned or sterilized with little difficulty — hive tools, smokers, protective clothing, extracting equipment, and the like.

The danger of disease also exists when you buy an established colony but there at least you have brood and comb that you can inspect. If you don't know what to look for, take someone with you who does. Any colony that you wish to buy should be inspected as a condition of sale, preferably by the state apiary inspector. If the seller produces a certificate of inspection, be sure that it is of recent enough date to be meaningful. It should be a certificate from the current season, and preferably reflecting an inspection within the past thirty days.

Mites, of course, add another dimension to disease inspection. With equipment that has not housed bees for a while, mites are not a problem. They do not survive for long when not in the presence of bees. In an established colony, though, they can be a problem and the inspection must include a search for mites. With tracheal mites this means taking samples of bees and doing a laboratory analysis. As of this writing, tracheal mite analysis is still time consuming, laborious, and certainly not routine. The several methods for varroa mite surveying differ, with each giving results with a different level of confidence. Stay in touch with your state apiary inspectors or other authority so that you know the status of mites and mite control for your state.

CHAPTER 16

FEEDING

BEEKEEPERS HAVE MANY DIFFERENT ATTITUDES towards feeding bees. Some do not believe in feeding established colonies, ever. Others feel that feeding is a negative reflection on their beekeeping ability. Some of these latter never feed their colonies; others may feed but won't admit it. This attitude is unfortunate. Not all colonies need to be fed, nor is feeding necessary every year. However, there are times when even the most competent of beekeepers must feed if the bees are to survive. There have been years when the spring weather was so poor that without being fed the colony could not possibly build up its population for a successful summer. There are years when insufficient nectar is available and the colony is not able to put enough away to get through the following winter. If the beekeeper does not intervene he or she may shortly become a former beekeeper.

Feeding is appropriate in any season if the bees need help. It is also appropriate whenever a new colony is established, whether it be from a swarm, a nuc, a package, or a split. Such new colonies are usually started during the period when established colonies are approaching or at the peak of population, ready to do their best work of the year. These new colonies are under strength, probably having little or no stores, and often, no drawn comb. Left to their own devices most of these colonies will probably become established, but a substantial number of them will be of questionable strength, will not make any surplus honey that year, and may not survive their first winter.

Routine feeding is most often done with syrup, presented in one of the standard top or internal feeders. (Winter feeding techniques are discussed in Chapter 1.) Be sure your feeder is large enough: feed generously. Think in gallons of syrup, not quarts. Depending on the particular time of year and the particular season, several gallons of feed may be in order. Feed if there is the slightest question. You will find your return in a better honey crop. However, do not leave honey supers on the hive if you are feeding. Take no chances of having honey contaminated with syrup.

Common beekeeper wisdom says to vary the proportion of sugar to water according to the time of year when mixing sugar syrup. Usually a 1:1 ratio is used in the spring and 2:1 (sugar:water) in the fall. Another way to look at this is to consider your goals at each time of year. In the spring a thinner syrup simulates a nectar flow and can stimulate the bees to a higher level of brood rearing. In the fall you are more interested in building up winter stores, and a thicker syrup is more efficiently processed by the bees.

Precautions in Feeding Bees

High-fructose corn syrup has become popular in recent years as a food for bees. It is not readily available in small quantities, however, which puts it out of the reach of most hobbyist beekeepers. Periodically there are reports of surplus HF syrup being available from sources such as soft drink or candy manufacturers. Beekeepers have been able to pick up pails or drums of this surplus syrup at attractive prices. The drawback here is that the syrup may not be pure — it may be syrup that has already been partway through the soft drink manufacturing process and has additives in it. These additives are usually toxic to the bees. Studies have shown that when fed this syrup the bees' lives are shortened significantly.

Not all such syrups are necessarily toxic, but it is difficult for the average beekeeper to know. The best practice remains — feed only pure granulated sugar or pure high-fructose corn syrup. Do not take chances with syrup from unknown or questionable sources.

Beekeepers sometimes tell of having fed various questionable foods — candy manufacturing by-products, maple syrup, molasses, and so on — with no apparent ill effects. It is difficult to know what effect a given food source is having on the bees. The bees probably would not die immediately, and how can the average beekeeper really tell if the life span of a bee has been shortened?

Aside from the possible toxicity of some potential feeds, there is the question of those impurities that do not poison the bees but that cannot be digested. These impurities pass on through the digestive tract to build up as wastes, waiting to be voided on the next suitable flight day. If the suitable day is long in coming, as so often happens in a long harsh winter, the bees have serious problems. This makes it even more important not to feed questionable materials in the fall, especially in the more northern climes.

Sugar Candy

Bees are fed most often with syrup or honey. However, sugar candy is an acceptable alternative in an emergency situation. In the middle of winter when temperatures and food reserves are low and something has to be done, mix some candy.

The ingredients are table-grade sugar and water. Some older books recommend adding tartaric acid to invert the sugar. This is no longer recommended. There is some evidence that tartaric acid shortens a bee's life.

To mix a five-pound batch, bring about two quarts of water to a boil in a medium to large pot. Turn off the heat and pour in the sugar, stirring until it is completely dissolved. At all times be careful not to burn or scorch the sugar. When it is completely dissolved, turn on the heat and bring back to a boil. Continue boiling, stirring often, until the mixture reaches the hard ball candy stage (260°-270°F). It is best to use a candy thermometer. This is a time-consuming process. It will take at least 30 to 40 minutes. When the mixture has reached the proper temperature, pour it out on sheets of waxed paper on a flat, hard surface. If you are working on a finished surface (the dining room table, for instance) put a thick layer of newspaper under the waxed paper for insulation: otherwise the heat may raise the finish. Raise the edges of the waxed paper with frame bars or similar sized sticks to keep the candy from running off.

When set, the candy will be hard, somewhat brittle, and of a light amber color. Suitably sized pieces may then be laid on top of the inner cover or directly on the top bars of the frames. The closer to the cluster it is placed the better.

An alternative method is to pour the hot mixture directly into a spare inner cover, rim up. The hole in the center of the cover should be kept open. The candy will adhere to the inner cover as it hardens, and the cover may then be placed on the hive rim down, replacing the regular inner cover.

Sugar candy is more difficult for the bees to use than is syrup or honey but it is an acceptable method of feeding. Again, be very careful not to scorch or burn the syrup. The resulting candy could harm the bees.

CHAPTER 17

POLLINATION

W̲ITH THE PROBLEMS BESETTING BEEKEEPERS TODAY, such as the spread of mites and the threat of Africanized bees, there are increasing restrictions on the interstate movement of bees. This threatens the continued availability of large numbers of colonies from the large, migratory operators. Pollination services provided by small, local beekeepers are becoming of more interest to both beekeepers and growers. Perhaps the biggest drawback to local pollination is that most beekeepers are hobbyists who individually do not have enough hives to satisfy the demands of the growers. A second problem is the quality of available colonies.

It is quite possible for several hobbyist or sideline beekeepers, each with only a few hives, to group their holdings to satisfy the needs of a particular grower. The logistics of such an endeavor may be difficult, but it can be done. The larger problem is in fulfilling the commitment with the requisite number of hives of proper quality and configuration. Pollination contracts usually are arranged several months in advance. No grower wants to be out looking for a beekeeper a couple of weeks before bloom. Any beekeepers, therefore, who are a party to a pollination agreement must be competent enough to bring through winter the number of hives committed, and the hives must be of adequate strength to do the job at hand.

This brings us to the heart of the matter. All hives are not suitable for pollination. There is a preferred configuration and a minimum size to ensure that the bees will do an adequate job of pollinating. The size of a colony has several facets, expressed in terms of bees, brood, food reserves, and the hive itself. Taken in order, a hive that is suitable for pollination:

■ will have a population in excess of 30,000 bees. A lesser population requires that a larger percentage of its adult bees tend brood. It thus cannot send out a sufficient field force to do the pollinating job at hand. As the population of the colony increases, so does the percentage of foragers.

■ will have at least six frames with brood. The amount of brood is a reflection of the size of the adult population. A normal, healthy colony will have this amount of brood in the late spring if its adult population is at least 30,000 adults. Note that this is six frames with brood, not six frames full of brood. This is in keeping with the spherical shape of the brood nest, where the outer frames are less than full.

■ will be housed in two deep bodies (or equivalent space). A colony of 30,000 bees or more requires this amount of space for normal development. Anything less may inhibit normal growth of the colony which may in turn reduce the amount of foraging and consequently the amount of pollination. A smaller hive may also encourage swarming and the consequent loss of half the population. Large colonies can also create problems. A colony in two hive bodies but with a large population, say 50,000 and up, may also swarm, and a hive with more than two bodies is more difficult to lift and move.

■ will have honey reserves in proportion to the population. Brood rearing and colony expansion depend on a certain level of food reserves present in the hive — three frames of honey, at minimum. Anything less may result in a reduction in brood rearing. A good honey reserve also allows the colony to expend a proportionally larger amount of its effort on pollen foraging. Though the bees accomplish pollination while foraging for either nectar or pollen, they are usually more efficient pollinators when foraging for pollen.

Some beekeepers have taken the attitude that if the available colonies are not strong enough, then add more. However, two weak colonies do not equal one strong colony. It has been well established that, for instance, two colonies of 15,000 bees each are not the equivalent of one colony of 30,000. As has been stated, on a percentage basis smaller colonies put out smaller field forces. With bee colonies, you can not make up for poor quality with additional quantity.

Pollination and Package Bees

Attempts have been made to pollinate with hives started from packages. It does not work. Referring to Figure 2-2, a colony started in late April from a three-pound package has not attained the adult population level required at the time that apples, blueberries, or even cranberries normally come into bloom. A three-pound package has an initial population of about 10,000 adult bees. By its nature, a package has no brood when first installed. The queen will start laying almost immediately but it will be more than three weeks before the first new bees emerge. Meanwhile, the original adult population is dying off normally. The result is a diminishing population during the first three to four weeks after installation. The requisite population of 30,000 bees is not reached until well into July and perhaps beyond, depending on circumstances of weather, nectar flows, and location. Using a larger package improves the picture somewhat but not enough to allow us to consider a package colony as a viable pollinating unit.

Is Pollination for You?

Contracting of hives to a grower for pollination, whether it be written or verbal, is a serious commitment. It is not something to be entered into lightly and unless it is done on a commercial scale it is likely to be in conflict with what might be your primary reason for keeping bees — honey production. Crops requiring pollination usually bloom in the May to June period, just when colonies are building up to maximum levels. In many areas honey production is at a peak. This is not a time to put stress on hives.

Pollination requires that the hives be moved twice during this period, and more often if more than one crop is involved. Each move is a setback to colony development and production. Furthermore, hives used for pollination are usually not supered for reasons of weight and ease of transportation. Storage space in the hive is therefore limited and the possible congestion that results may encourage swarming. Surplus honey from a hive used for pollination becomes a chancy thing.

Pollination is hard work. Orchards, cranberry bogs, and blueberry patches are often difficult of access, especially at night when most of the moving is done. And even without supers, hives are heavy and awkward. Your friends and your spouse always seem to have pressing business elsewhere at the time that the hives need to be moved.

If you think you would like to rent your hives out for pollination, start on a small scale. Work into it slowly and decide if it is really what you want.

APPENDIX A

DISEASES, MITES, AND AFRICANIZED BEES

■

Every beekeeper should be able to recognize all of the common bee diseases, and should also know what to do about them. A discussion of the various symptoms and treatments is not within the scope of this book, however. Some excellent books on the subject are readily available, as are booklets and bulletins from various state and federal agencies. Some of these books are listed in Appendix B. The Cooperative Extension offices in many states also have material available, either free or at nominal cost.

Mites are a separate problem. At the time of this writing both tracheal and varroa mites are found in many parts of this country and they are spreading. At the time of your reading they may very well be everywhere. Tracheal mites were the first arrival, being found in the U.S. in 1984. Varroa mites were discovered here in 1987. Both kinds of mites are creating a serious problem. With the arrival of the varroa mite, tracheal mites, at least temporarily, took a back seat. They are still a very serious problem, however, and in no way can they be ignored.

Distribution of mites, our knowledge of them, and our methods of coping with them are all still evolving. Beekeepers should be in touch with their local and state beekeeping organizations and with appropriate local and state agencies in order to know the very latest in detection and treatment methods.

Tracheal Mites

The tracheal mite is an internal parasite of honey bees. It lives and breeds in the bees' tracheal tubes, puncturing the walls of the tubes to feed on hemolymph. Individual bees' lives are shortened, and in heavy infestations, entire colonies die. There is no conclusive way to know if you have tracheal mites in your hive other than through laboratory dissection and analysis of a significant number of bees. This is a tedious and time-consuming procedure and only a few facilities offer this diagnostic service.

We have a great deal to learn about tracheal mites before we bring them under control. Right now there is a considerable amount of confusion. Some of the things we do know are:

■ There are geographical differences between mites, from one country to another and apparently within the U.S. This is a contributing factor to the confusion we are experiencing as to effects and treatments of mites.

■ Mite populations are cyclical. There is a fall buildup, a winter peak, and a summer crash. This cycle reflects an opposite cycle in the bees' population. A bee colony peaks in summer, tapers down in the fall, and hits a low point in the winter. A good queen can easily outproduce the mites in the late spring and early summer so that the percentage of infested bees is lower and the colony as a whole is healthier. The opposite happens in the fall and winter when the queen is laying from few to no eggs, allowing the mites to get ahead. Spring requeening will take advantage of this cycling.

■ Menthol is the only treatment now available against tracheal mites, and the results are erratic. Some negative reports probably stem from improper use; for it to be effective, the weather must be warm. Menthol, which is available to us in a crystalline form, must vaporize to be effective in the hive. The resulting gas circulates through the hive and kills the mites. The minimum ambient temperature for this is 60°F. At lower temperatures menthol does not vaporize. At temperatures between 60° and 80°, the menthol should be placed above the brood nest. Above 80°, it should be placed on the bottom board. Otherwise, the bees will be driven from the hive by the fumes. At temperatures starting around 95°, menthol will melt and run.

■ The life cycle of an individual mite runs fourteen days. This is from the time an egg hatches, matures to become an adult, mates, and lays eggs of its own. Menthol kills only the adult mites. To be effective then, menthol must be in the hive, and vaporizing, for fourteen consecutive warm days. Otherwise, some adult mites will not be killed and will go on to perpetuate the infestation. If cool weather intervenes, the treatment must be continued.

■ Even after a laboratory report indicates that a colony of bees is mite-free, the only positive conclusion that we can draw is that the sample tested was mite-free. A certificate stating that a colony was free of tracheal mites is not a guarantee. It simply indicates that within the parameters of the particular sampling technique for that test, the colony appeared to be mite-free. The only way to be 100 percent sure is to kill and examine all of the bees of that colony. To know the real value of a particular certificate, you must know the number of bees collected for the test, the number actually tested, where in the hive they were collected, how long ago the test was made, and where the hive has been and what exposure it has had since the sample was taken. You must also know the definition of mite-free for that particular test. Some testing in the past has allowed for a colony to be declared mite-free if there are less than a certain minimum of infested bees found—five infested bees per 100 bees examined, for instance.

If you suspect tracheal mites in your hive, contact your state apiary inspector. If testing facilities are available, the inspector can tell you how to collect and preserve bee samples properly, and how and where to send them.

Varroa Mites

Varroa mites are external parasites of honey bees and are visible to the naked eye. They are found in brood cells with immature bees, and on the bodies of adult bees. In both instances they puncture the bees' body wall to feed on hemolymph. Heavily infested brood will not develop properly and may not successfully emerge as adults. Those adults that do emerge will be weakened, and their lives may be severely shortened.

Varroa mites are found in many states and are spreading. Ultimately, they will be everywhere. Inspection for the presence of these mites should become a part of your hive management routine. Even if they are not found in your area now, assume they will be and be prepared.

Although varroa mites can be seen with the naked eye, they are not seen easily. In fact, even with several mites present on a single bee, they are easily missed. When not actively feeding, they hide on the bees' bodies, in the space between the thorax and the abdomen, for instance. As might be expected, low levels of mite infestation are harder to detect than high levels. Low levels will grow to become high, however. An infested colony that is not treated can be expected to die within three to five years. Some relatively simple techniques for inspecting are:

■ ETHER ROLL. *Equipment:* a round glass jar, about 10- to 16-ounce capacity, with a tightly fitting cover, and a can of automobile starting fluid (ether). Collect about one half a jar of live bees. Open the jar carefully and give the bees a one second squirt of ether. Recap, then hold the jar on its side and roll it for about one minute to make the bees tumble. A film of moisture will form on the inside of the glass. Any mites present on the bees will fall off and will stick to the moist glass, and so be obvious. After observing, release the bees quickly and they may revive.

■ CAPPING SCRATCHER. *Equipment:* a capping scratcher. Mites lay their eggs in both worker and drone cells, though drone cells are preferred. In the hive find an area of capped drone brood. Use the capping scratcher to lift off a section of cappings, along with the underlying drone pupae. Look carefully at the pupae for the presence of mites. It is not enough to examine just a few cells. To be effective, several hundred drone cells must be examined. In the past, beekeepers tended to keep down the amount of drone brood in a hive. Now, to enhance inspection for mites, many beekeepers are encouraging drone brood by installing drone-sized foundation.

■ SHAKING. *Equipment:* alcohol, detergent, gasoline, or other similar material, and a container with a lid. Dead bees (several hundred, if possible) may be checked by placing them in the container with the liquid and shaking vigorously for about a minute. Any mites present should fall off. Strain or sieve the contents and look carefully for mites.

The fact that no mites are found using any of the foregoing techniques does not mean that no mites are present. Again, low levels of infestation are difficult to detect. Keep checking routinely. Check often if you are in an area where varroa is known to be present.

If mites are detected, then treatment is in order. At this time, only one treatment material is available—Apistan™ (fluvalinate). Apistan™ is a pesticide, registered and controlled as such by the EPA. Until recently it has been available only to qualified personnel, licensed pesticide applicators, for instance. As of mid-1990, Apistan™ has received a Section 3 registration, making it readily available to all beekeepers.

Unfortunately, Apistan™ is not intended as a routine preventive. It is to be used only after varroa has been actually detected in a hive. As with any chemical treatment put into a hive, great care must be taken to understand and follow the labelling instructions.

Africanized Bees

In recent years there have been a number of reports of the Africanized bee entering the United States, most often as swarms carried unknowingly on ships from South or Central America. As far as we know each of these swarms was destroyed at the time it was detected. However, it is reasonable to believe that not every such swarm was detected, and that some small number of them may have become established in this country. If this has indeed happened, we have not yet seen any seriously adverse effects.

In the fall of 1990 the first Africanized bee swarm to come overland into the United States entered Texas from Mexico. This event was met with surprising calm by the media and the populace — surprising because of the excesses of publicity that have greeted some of the more casual incursions of this bee in the past. The swarm was destroyed and, as the result of a careful survey, no other Africanized bees were found in the area. More will come, however.

As of this writing (late 1990) we still cannot predict accurately the spread or behavior of the Africanized bee in the United States. Reports from South and Central America are replete with inconsistencies and conflicting information. They are good honey producers; they are poor honey producers. They are good pollinators; they are poor pollinators. They will not survive in cold climates; they can survive as far north as Minnesota. Only by living with them and experiencing them can we know what they will be like. As they spread through whatever part of the United States they are able to inhabit, we can expect continued confusion. These bees do act inconsistently, probably for several reasons. Geographical factors such as latitude and altitude above sea level, degree of hybridization, size of colonies, and forage availability all play their part. It will no doubt be years before we have a reasonable understanding of these bees in our country.

One consistent message that seems to come through from those who live with the Africanized bee is that living with them is possible. Some South American beekeepers go a step further: they believe that the Africanized bee is a better bee and that they are better off now than when they kept only European bees. The Africanized bee does create some problems, though, and adjustments will be required. We will become more careful in locating bee yards, and in our hive management techniques. There may be fewer colonies, and municipal and state governments may impose various restrictions. However, honey bees will continue to be an important part of our lives.

Finally, we must remember that aggressive colonies of European bees do exist. We must keep reminding ourselves and everyone else that excessive stinging does not automatically mean that a colony is Africanized.

APPENDIX B

ADDITIONAL READING

▬

MANY GOOD BOOKS HAVE BEEN WRITTEN ABOUT BEES and beekeeping, and new ones are appearing regularly. The array can be bewildering. Some books are outstanding, others less so. An experienced beekeeper usually has the basic knowledge to pick and choose, from both the vast number of books available and the contents of individual books. The less experienced beekeeper may benefit from some guidance. Following is a list of books that I believe to be worthwhile. They differ in their approach and level of detail but all have something to offer. This list is not intended to be all inclusive. There are many other good books. This list is but a starting place.

General Reference

Dadant and Sons, ed., *The Hive and the Honey Bee*, rev. ed., Dadant and Sons, Hamilton, Ill., 1974.

Morse, Roger A., and Ted Hooper, ed., *The Illustrated Encyclopedia of Beekeeping*, E.P. Dutton Inc., New York, 1985.

Root, A.I. *The ABC and XYZ of Bee Culture*, 40th ed., A.I. Root Co., Medina, Ohio, 44256, 1990.

Practical Beekeeping

Morse, R.A., *The Complete Guide to Beekeeping*, 3rd ed., E.P. Dutton, New York, 1986.

Sammataro, Diana, and Alphonse Avitabile, *The Beekeeper's Handbook*, 2nd ed., Macmillan, New York, 1986.

Taylor, Richard, *The How-To-Do-It Book of Beekeeping*, 3rd edition, Linden Books, Interlaken, N.Y., 1980.

Biology, Organization and Communications

Free, John B., *The Social Organization of Honeybees*, Edward Arnold Ltd., London, 1977.

Frisch, Karl von, *Bees: Their Vision, Chemical Sense, and Language*, rev. ed., Cornell University Press, Ithaca, N.Y., 1971.

Seeley, Thomas D., *Honeybee Ecology*, Princeton University Press, Princeton, N.J., 1985.

Winston, Mark, *The Biology of the Honey Bee*, Harvard University Press, Cambridge, Mass., 1987.

Pollination

Free, John B., *Insect Pollination of Crops*, Academic Press, New York, 1970.

McGregor, S.E., *Insect Pollination of Cultivated Crops*, U.S. Dept. of Agriculture, Washington, 1976.

Diseases

Bailey, Leslie, *Honey Bee Pathology*, Academic Press, New York, 1981.

Hansen, Henrik, *Honey Bee Brood Diseases*, Wicwas Press, Ithaca, N.Y.

Morse, Roger A., ed., *Honey Bee Pests, Predators, and Diseases*, Cornell University Press, Ithaca, N.Y., 1978.

Historical Interest

Langstroth, L.L., *Langstroth on the Hive and the Honey Bee*, A.I. Root Co., Medina, Ohio, 1853, reprinted 1977.

Miller, C.C., *Fifty Years Among the Bees*, Molly Yes Press, New Berlin, N.Y. 1915, reprinted 1980.

INDEX

Numbers in boldface indicate photographs or illustrations.

Queen cups, 20
Queen excluder, 27-28, 85, 106
Queenlessness, 3, 76-77
Queen maturation time, **36**
Queenright colony, 35

R

Requeening, 9, 24, 35-37, 77-78, 96, 100
Reversing, 10, 24
    frequency, 11
Robber bees, 2-3
Robbing, 47-48, 70-71
    from wet supers, 107-8
Round comb, 83-84

S

Scouts, 21-23
Section comb, 84
Shifting frames of honey, 24
Skunks, 71-**72**
Slatted rack, 117, 122-23
Snelgrove board, 27-**28**
Spacers, 63, **64**, 65
Starvation, 3
Stinging, 6
Storage space, 73
Strains of bees, 34
Sugar, 5, 133-34
Sugar candy, 5, 133-34
Sugar syrup, for feeding, 4, 132-33
Supering, 25, 28, 57-65, 73, 79, 85, 97
    contents, 58
    definition, 57
    in fall, 99-100
    number of frames, 63-65
    and placement, 60-61
    removing supers, 108-112
        blower, 110-**112**
        brushing, 109
        burr comb, 108
        escape board, 110
        fume board, 110, **111**
    size of supers, 61-63
    strategies, 61
    and timing, 60
    wet supers, 107-8
    in winter, 105-6

Supersedure. *See* Requeening
Supply and demand, 86-87
Swarming, 14, 17-30, 69, 83, 85
    capturing, 26, 29
    and clipping, 39-40
    and colony size, 19
    and congestion, 19, 58
    control, 23-25
    departure, 21-23
    to perpetuate the species, 17-18
    preparation, 20-21
    prevention, 25-28
    and the queen, 19
    removing frames of brood, 26
    season, 18
    splitting a hive, 26, **27**
    swapping hives, 26-27
    and worker age, 20
Swarming urge, 10, 23-24

T

Tartaric acid, 5
Terramycin, 43, 114
Timing, 10-11
Top feeder, 8
Traffic volume, 69-70

V

Ventilating, 56
Ventilation, 4, 117, 119-20
von Frisch, Karl, 74

W

Wax, 59, 81, 82, 96
Wax glands, 58, 61
Wax moths, 60
Weather, 5-6
Weight checks, 68-69
Winterkill, 2-4
    causes of, 3
Worker, 51-56
    duties, 54
    phases, 51-53
Worker age, and swarming, 20
Workers, laying, 77-78
Wrapping, 119-20